CHRONOLOGY O
EARTH AND THE GREAT CONTROVERSY

RELATION TO CHRIST'S LAST GENERATION PROPHECY AND WHEN HE WILL RETURN

CHIMEREMEZE NWORGU

6305060406

Kingpriest777777@gmail.com

2417 FLEETWOOD DRIVE, JOLIET IL 60432

A STRAIGHT TRUTH AND TESTIMONIES MINISTRY PUBLICATION

LCCN: 2025948644

TABLE OF CONTENTS

GENERAL INTRODUCTION AND INSPIRATION, BASIS, AND GOAL OF THIS STUDY

Biblical chronology is not a very appealing and exciting subject, in addition to being seemingly non-salvific. But it is a subject Jesus was much aware of its importance, as is evident in his declaration regarding the latest and last generation that will not pass before the end-time events he prophesied are fulfilled. The necessity and basis of this study is therefore the understanding of Jesus' ***LAST GENERATIONS PROPHECY*** which is documented in Matthew 24 and the other related books of Mark and Luke. It is to identify exactly when the last generations of the Jewish nation and the world, respectively, began, and the very events that mark or dot these periods. A ***GENERATION*** refers to a specific period of time, and therefore has a point in time that marks its beginning, as well as how long it lasts. When Jesus gave the prophecies, he was addressing an audience of various ages, from infants, children, teens, young adults, old adults to the elderly. Each of these age groups belong to various generations based on when they were born and based on their own distinct genealogies. But Jesus specifically used the word in the singular, ***"THIS GENERATION…"***, an indication that he meant a particular generation and had a particular definition of the length of a generation.

So then, there are two aspects to a generation, namely, genealogy and length of a generation. There is also the length of time for generation turnover, that is, how long before a new generation comes up. In a genealogy, there are many generations depending on how many progenitors that make up the genealogy. Thus, we see in Matthew 1, Jesus's genealogy from Abraham numbering 42 generations. While the length of a generation is the same or constant, the age at which a new generation is born varies.

Thus for the 42 generations, the age at which each generation was born by their parents differ from each other. A new generation is born before the generation of the parent passes, ends, or concludes. Since Jesus was addressing an audience of different ages, for them to belong to the same generation that he was referring to, he must base this generation on a parent that they all have in common, and upon a specific genealogy that runs from this parent. What particular generation would a Jewish audience of various ages belong to and have in common, and what genealogy has an unbroken line from this parent to Christ's time? The answer is that they all have the same ancestor from whom they originated, and that ancestor is Abraham. Therefore, beginning with Abraham, down to the time Jesus was making the statement to them, they all belonged and were existing in ***THE LATEST GENERATION FROM ABRAHAM***.

Thus Christ meant the ***LAST GENERATION OF THE NATION.*** For this, Christ had to pick a lineage that began from Abraham to the present day he was talking to them. Matthew 1 tells us what lineage that Christ based his last generation on, his own lineage from Abraham. This does not mean that everyone in the audience will live to the end of this generation in question. Some of his audience died that very year. Some lived until the very day the nation was destroyed in AD 70. But that generation Jesus was referencing has a determined length of time that it will normally last. What Christ meant therefore was that that latest generation of his lineage from Abraham will not complete its standard length or period before the nation will be destroyed by Rome, the Abomination of desolation. The last generation is therefore simply the latest from Abraham, and the reason it is also the last is because it will not run its course before it is ended, In other words, it will end prematurely.

This is what Jesus was addressing, the latest and present generation of the people since and from Abraham to that time he was giving the prophecy to them. How would Jesus know the latest and present generation that he was directing the prophecy to, and that the generation will not run its course? The answer is firstly ***BIBLICAL GENEALOGY AND CHRONOLOGY***, and secondly, knowledge of the standard length of a generation(not the same as the age when a new generation begins). This means that Jesus did study the genealogy and chronology of the world and especially that of Israel, which starts from Abraham the father of the nation. It is through this study that Jesus learnt that he belongs to the latest of the generation of the Jewish race from Abraham. It is this knowledge and the certainty of it which Jesus had that Matthew recorded in Matthew 1:1-18. While Matthew 1 genealogy gave Jesus the number of progenitors from Abraham to himself, 42, of which he was the latest, it does not give him the length of a generation. The much he can get here is the average length of a ***GENERATION TURNOVER***. Rather the interest is in how many progenitors there are in the lineage, 42 altogether in this case, from Abraham to Jesus(Matthew 1:17).

In contrast, the length of period that any given or particular generation lives before dying off is found in Genesis 15:13-16 and Psalm 90:10. In his prophecy about the LATEST AND LAST GENERATION that shall not pass, Christ was referencing both Matthew genealogy(that tells him that he is the latest generation from Abraham) as well as to the chronology(number of years from Abraham to his birth. So, Jesus trusted and knew that biblical chronology can be accurately determined, and that there is no mistake or gaps in it, all the way from Adam to his birth. This is very crucial because it gives us confidence that if we calculate the chronology of the earth accurately, based on scriptural data, we will know how old the world is.

In other words, Jesus was telling the people that the latest generation from Abraham the father of the nation, a generation that began with his birth, will not complete its normal length before all the things he prophesied come to pass, which in the case of the Jewish nation culminates in the destruction of the city and national ruin. It is this latest generation of the people from Abraham that Jesus was addressing when he made the declaration, **"Verily I say unto you, this generation shall not pass, till all these things be fulfilled"** – Matt 24:34. The only way Christ could know that this generation will not run its course or length of time can only be from knowing the standard length of a generation. This knowledge is provided in scriptures by Psalms 90:10 and Genesis 15:13-16, which can both be verified with, and corroborated by, actual calculation regarding the 430-year sojourn of the people in Egypt. Let us therefore proceed to undertake this interesting though challenging task.

Along the line of this study, the following questions would be addressed: *IS THE 7000 YEARS REDEMPTION CALENDAR THEOLOGY BIBLICAL OR AN OVERREACH? WHAT ACTUALLY IS THE MESSAGE OF THE 4000, 6000 YEARS BY ELLEN G. WHITE?*

THE LENGTH OF A GENERATION

The most precise definition of the length of generation known in scriptures, is that provided by David in Psalms 90:10, and that found in God's prophetic promise to Abraham regarding how long his grandchildren will be in Egypt and when and how He will deliver them - Genesis 15:13-16.

"The days of our years are three score years and ten; and if by reason of strength they be four score years, yet is their strength labour and sorrow; for it is soon cut off, and we fly way".

After Jacob, the human life span fell drastically to what David indicated in the above psalm. This is the very life span, or length of generation that Jesus was referencing in his last generation prophecy- seventy to eighty years. Taking the maximum range, which is eighty, we can therefore paraphrase Jessus's prophecy to the people as follows:

"Verily, verily I say unto you, this latest generation from Abraham, which is my generation(See Matthew 1) shall not reach its summit of eighty years before this city and this nation will be destroyed by the people of the Prince of Rome, the Abomination of Desolation"-Matthew 24: 34 (see Daniel 9:27)

As we study the chronology of the earth and the Great controversy in more detail, it will be seen that this prophecy fulfilled exactly in accordance with Jesus's specifications. Jesus was born in 2 BC, and Jerusalem and the nation of Israel was destroyed in AD 70, a length of only 72 years, short of the eighty years span of a generation. This shows that we are correct in taking the maximum of the 70 – 80 years range that David gave in Psalms 90:10.

Now let us see if God's words to Abraham regarding how long Israel will be afflicted in Egypt will corroborate David about the length of a generation.

"And he said unto Abram, Know of a surety that thy seed shall be a stranger in a land that is not theirs, and shall serve them; and they shall afflict them *four hundred years*; [14] And also that nation, whom they shall serve, will I judge: and afterward shall they come out with great substance. [15] And thou shalt go to thy fathers in peace; thou shalt be buried in a good old age. [16] *But in the fourth generation* they shall come hither again: for the iniquity of the Amorites is not yet full" - Genesis 15"13-16.

By the fourth generation God clearly meant the fourth from after Joseph's death in Egypt, because the affliction of the people did not start until Joseph and his brothers had died(See Exodus 1:6-8). Was God being exact or was He being approximate? The answer to this question will reveal the length of time God assigned to a generation from that point forward. We can only get the answer by determining when Joseph died relative to the overall number of years of the sojourn in Egypt. This fact can only be gotten from studying and calculating the chronology of the earth, especially from Abraham to Exodus. While the detailed chronology of the earth will be provided later in the study, let us at this juncture provide the reader a snapshot that focuses on Abraham to the Exodus.

SNAPSHOTS FROM THE CHRONOLOGY OF THE EARTH: ABRAHAM TO EXODUS

ABRAHAM TO EXODUS: **100 + 60 + 130 + 430 = 720 YEARS**

Abraham was 100 years when he begat Isaac.

Isaac was 60 years when he begat Jacob and Esau(twin brothers).

Jacob(Israel) was 130 years when he, along with his other sons and his grandchildren, began his sojourn in Egypt. Joseph was already in Egypt.

Ussher was wrong in reckoning the 430 years of sojourn from when Abraham left Haran. Generations from Abraham to exodus is well more than four generations. In fact both based on Matthew 1, as well as Psalm 90:10, Ussher's view is in complete disharmony with the Genesis 15:13-16. Abraham's stay in Egypt was so abrupt and too abrupt to count as a sojourn(See Genesis). Moreso, it is his seed that God said will sojourn in Egypt. Thirdly, Isaac never set foot in Egypt, as God warned him(See Genesis). Therefore contrary to Ussher, the correct reckoning of the 430 years begins when Jacob, at age 130, entered Egypt at joseph's invitation(See Genesis).

With Ussher's view corrected, let us now get into the actual number of years that the people of Israel were afflicted. While God said four hundred years, it will be seen that God was simply being approximate. We will see that God, who is all wise, knowing, and Almighty, is not ignorant of the rules of approximation, because after all He is the creator of the laws of nature which includes the laws of physics and mathematics.

ABRAHAM TO DEATH OF JOSEPH: 100 + 60 + 130 + 78 = **368 YEARS**

LENGTH OF AFFLICTION (ASSUMING THIS STARTED THE SAME YEAR JOSEPH DIED):

430 – 71 = 359 YEARS.

CITED SOURCES AND RESOURCES:

Jacob entered Egypt at the age of 130 years old (Genesis 46:1-4; 47:8-9).

Joseph was 17 years old when he entered Egypt (Genesis 37:2).

Joseph was 30 years old when he became Governor of Egypt (Genesis 41:1-46. Emphasis on verse 46).

The famine was 2 years old when Jacob and his other children entered Egypt (Genesis 45:5,6).

Therefore Joseph was 39 years old when Jacob and his brothers came to Egypt (7 years of plenty + 2 years of famine + 30 years he was when he became Governor). This is his age when the 430 years of Israel's sojourn began.

Joseph died at the age of 110 years old (Genesis 50:26). This means that he lived 71 years more after the 430 years of sojourn began. This means that the actual affliction and suffering of the people lasted 430 – 71 = 359 YEARS.

If Joseph went to prison the very year he went into Egypt, this means he was in prison for 13 years (His faith in God and in the dreams he received from God gave him hope and strength).

So, God was being approximate. This is very important because we can understand why God also gave Ellen G. White the 4000, 6000 years prophecy in approximation. We will discuss the Ellen G. White 4000, 6000 years prophecy later in the study. In conclusion therefore, we have corroboration and confirmation from Genesis 15:13-16 that David was correct in giving the length of a generation as 80 years-Psalms 90:10.

If this definition of the length of a generation is accurate, and Jesus's birth date marked the beginning of the last generation of the Jewish nation, then we will have this harmonizing with accurate chronology of the earth from Adam to Jesus Christ. In other words, chronology will show that Christ was born at the very end of the latest generation from Abraham, and thus his birth date marking the very beginning of the next and last generation that was not to pass before Jerusalem was destroyed by Rome. Here is what chronology, as accurately computed in this study, shows(the details of the study will be given shortly after in the book):

<u>CHRONOLOGY FROM ADAM TO JESUS:</u>

2728(ADAM TO EXODUS) + 479 + 37 + 390 + 527 - 2 = 4159 YEARS.

With the length of a generation being 80 years, we will have:

80 years dividing 4159 years = 51.9875

THE FRACTION 0.9875 IS THE NUMBER OF YEARS INTO THE 52ND GENERATION THAT CHRIST WAS BORN = (0.9875 X 80) = 79 YEARS.

So Christ was born at the very end of the 52ND GENERATION, meaning that the birth of Jesus marked the beginning of the 53rd generation from Adam. This is therefore the generation that Jesus meant it will not pass, i.e. complete its length of 80 years before Rome destroys the city. Sure enough the city was destroyed when this 53rd generation was only 72 years old.

NOTE: While Matthew 1 provides Christ's genealogy from Abraham, it is not a chronology because it does not give the ages of the progenitors in the lineage. Chronology up to Christ's birth is provided from Adam. The relevant texts are provided further down in the study under the detailed chronology of the earth.

LENGTH OF A GENERATION IS NOT FORTY YEARS THIS TEACHING IS INCORRECT

The theology that defines a generation as 40 years long is based on faulty interpretation of God's words regarding the disobedient generation that left Egypt Canaan. If anything it is about the average age that a progenitor gives birth to the next generation-the ***GENERATION TURNOVER AGE.*** This turnover age ranges from 40 – 51 years. We get 51 years as an average generation turn over age when we divide the 2151 chronology years from Abraham to Jesus with Matthew 1's 42 generations. This range therefore is what is called the TURNOVER AGE FOR GENERATIONS. It is a range and an average because progenitors in a lineage give birth to next generation at varying ages.

So it is in regard to this generation turnover age that Paul wrote about the forty years, and was not giving the length of a generation. It reads as follows:

Hebrews 3:7-19 KJVS

"Wherefore (as the Holy Ghost saith, To day if ye will hear his
voice, [8] Harden not your hearts, as in the provocation, in the day
of temptation in the wilderness: [9] ***When your fathers tempted me,
proved me, and saw my works forty years***. [10] ***Wherefore I was
grieved with that generation***, and said, They do alway err in their
heart; and they have not known my ways. [11] So I sware in my
wrath, They shall not enter into my rest.) [12] Take heed, brethren,
lest there be in any of you an evil heart of unbelief, in departing from
the living God. [13] But exhort one another daily, while it is called
To day; lest any of you be hardened through the deceitfulness of sin.
[14] For we are made partakers of Christ, if we hold the beginning

of our confidence stedfast unto the end; [15] While it is said, To day if ye will hear his voice, harden not your hearts, as in the provocation. [16] For some, when they had heard, did provoke: ***howbeit not all that came out of Egypt by Moses***. [17] ***But with whom was he grieved forty years? was it not with them that had sinned, whose carcases fell in the wilderness?*** [18] And to whom sware he that they should not enter into his rest, but to them that believed not? [19] So we see that they could not enter in because of unbelief".

Nowhere in the above texts is it derivable that a generation is 40 years long. Paul employed the term, generation, as it applies to a group of people that have one thing in common, namely, they all left Egypt together and they all committed the sin of lack of faith and rejection of God's redemption, preferring Egyptian bondage to their deliverance. To conclude that a generation is 40 years long is to say that only those who were born on the day the people left Egypt are the ones that died. But when God said, "***THAT GENERATION",*** He meant both young and old who left Egypt. It is therefore in the same manner that Christ employed the word in his Matthew 24 prophecy.

The only length of time given for a generation is therefore approximately 70-80 years, based on God's prophetic promise to Abraham and Psalmist. As we calculate the chronology of the earth from Abraham to the exodus, it will be discovered that it is exactly nine generations(720 / 80). So the ninth generation from Abraham just ended on the day they left Egypt. From combining this chronology(720 years) with Matthew 1(42 generation turnovers), we get the average generation turnover of 51 years. This is the average age which a progenitor is when he gives birth to the next generation. Based on average turnover(51 years) of generations, those who left Egypt were therefore at the most, about 40-51 years old, and they were the 10th generation from Abraham. This is the generation that God punished for forty years in the wilderness. As can be seen, in forty years most of that generation would die off based on the generation length of 80 years, thus confirming Psalm 90:10 definition of 80 years length for a generation. Of course there are exceptions, those that live beyond the age of 80. But God was determined that that generation die off except Caleb. Only Caleb lived beyond the 80 years length. Joshua was much younger than Caleb. Moses was especially blessed and would have lived beyond the 80 years also.

So, in that prophecy in Matthew 24, Jesus was essentially saying to them:

"***THIS LATEST GENERATION, MY OWN LINEAGE AND GENERATION FROM ABRAHAM THROUGH DAVID, WILL BE THE LAST GENERATION BEFORE THE NATION AND PEOPLE AND CITY ARE DESTROYED***". *IT WILL NOT COMPLETE ITS LENGTH OF 80 YEARS".*

This should have made sense to those who were discerning and good students of prophecy, because Daniel's prophecy says categorically that the nation was brought back because of Abraham and Christ alone, that is, based on the everlasting covenant God made with Christ and for which he made Abraham temporary custodian(See Daniel 9:25-The city and temple was built and restored unto the Messiah; Galatians 3:16-17). Since the nation was brought back and the city rebuilt and restored unto and for the Messiah, it stands to reason that the only way the nation could have continued to exist was if she had accepted Jesus as her Messiah and king. But sadly, they did the very opposite, preferring Barabbas and Caesar to their rightful king: "Give us Barabbas", and regarding Jesus, they all shouted, "Crucify him". In their blindness they TOOK AWAY THE DAILY(their savior and God and king), and chose Rome and Caesar(which Jesus identified as the Abomination of desolation).

THE LAST GENERATION OF THE WORLD- A PARALLEL OF CHRISTS PROPHECY ABOUT THE LAST GENERATION OF THE JEWISH NATION

If the chronological computation done in this study is accurate, then it will show that in the present year of AD 2025, the world is in its LATEST GENERATION, and if our interpretation of Daniel 12(presented later in the study) is accurate, it will also be the LAST GENERATION that will not pass(will not complete its 80-year lap) before the world is destroyed by the second coming of Jesus. So, let us find out.

CHRONOLOGY FROM ADAM TO AD 2025

ADAM(CREATION) TO EXODUS: 2728 YEARS

EXODUS TO KING SOLOMON'S FOURTH YEAR ON THE THRONE: 479 YEARS

THE REST OF SOLOMON'S REIGN: 37 YEARS(HE REIGNED A TOTAL OF 40 YEARS)

SOLOMON TO CARRYING INTO 70 YEARS CAPTIVITY: 390 YEARS

FROM CARRYING INTO CAPTIVITY TO AD 1: 527 YEARS.

ADD THE PRESENT YEAR OF 2025: 2025 YEARS

GRAND TOTAL(ADAM TO THE YEAR AD 2025) = 6186 YEARS

NO OF GENERATIONS(80 YEARS LONG) = 77.325 GENERATIONS(So we are 0.325 YEARS INTO THE 78TH (latest) GENERATION OF THE EARTH AS AT THE YEAR 2025).

0.325 0F 80 YEARS = 26 YEARS(THE LATEST GENERATION-78TH GENERATION IS 26 YEARS OLD) .

THEREFORE THE 77TH GENERATION OF THE EARTH ENDED: 2025 – 26 = AD 1999(The 77th generation ended on the last day of AD 1999).

THEREFORE THE LATEST GENERATION OF THE EARTH, THE 78TH GENERATION BEGAN ON JANUARY 1, 2000, AND IS ON ITS 26TH YEAR.

IF THEREFORE THIS 78TH GENERATION IS TO BE THE LAST GENERATION OF THIS EARTH, THE GENERATION THAT SHALL NOT PASS JUST AS THE LAST GENERATION OF THE JEWISH NATION, THEN THIS WORLD DOES NOT HAVE UP TO(80 – 26 = 54 YEARS).

But do we have evidence from Daniel 12, from current events that are gearing to fulfill Daniel 12, that indicate that this present generation has only until between the year AD 2038 and AD 2042?

THE LATEST GENERATION OF THE EARTH AS AT THE YEAR AD 2025 AND WHY WE KNOW FOR CERTAIN THAT IT IS THE LAST GENERATION OF THE WORLD

So to the question, Are we living in the latest 80-year long generation of the world?, the answer is yes. It began in the year, A.D. 2000, based on chronology from Adam. This present generation is therefore only about 26 years old(in its 26th year) as of this year of 2025. But how do we tell it will be the LAST GENERATION OF THE WORLD? The answer is, if we see the repetition of the prophetic events that occurred in the LAST GENERATION OF THE JEWISH NATION. What are these events? The important Great Controversy events of the last generation of the Jewish nation were as follows:

1. **2 BC - *THE BIRTH OF CHRIST. Marked the beginning of the Last Generation of the Jewish nation.***
2. **AD 27 – *THE BAPTISM OF CHRIST- The gathering of the saints into the kingdom and preaching of straight truths and testimonies by Jesus thus causing a SHAKING IN AND OF THE THEN CHURCH, the JEWISH CHURCH.***
3. **AD 31 - *THE TAKING AWAY OF THE DAILY* (GOD AND JESUS PLUS THEIR HOLY COVENANT AND HEAVENLY SANCTUARY MINISTRATIONS ARE**

THE DAILY. THEY ARE THE TRUTH PERSONIFIED AND WERE TAKEN AWAY AND CAST DOWN BY THE UNION OF APOSTATE CHURCH AND STATE LEADING TO HIS CRUCIFIXION(PHYSICAL REMOVAL FROM HIS NATION AND CHURCH) AND ABOMINATION OF DESOLATION(ROME AND CAESAR AND THE SUN GOD)-"We have no king but Caesar", "If thou let this man go, thou art not Caesar's friend", Crucify him", "Gove us Barabbas"

AD 34-*COMPLETION OF THE LAST OF THE SEVENTY WEEKS GIVEN TO THE NATION TO FILL THE CUP OF HER INIQUITY AND TRANSGRESSIONS(See Daniel 9). END OF PROBATION FOR THE NATION*-FINAL REJECTION OF CHRIST AS THE SPIRIT OF GOD in the STONING OF STEPHEN-34 A.D.

4. **AD 66 - *ABOMINATION OF DESOLATION* TAKES HIS STAND IN THE HOLY PLACE.**

5. **AD 70 - *THE END*-DESTRUCTION OF THE APOSTATE NATION AND PEOPLE.**

Did Christ address only the last generation of the Jewish nation-church, or did he give a prophecy that has dual application? Any careful student or scholar of Bible prophecy will not fail to notice that Christ's prophetic statements have dual application. Christ gave two PARALLEL LAST DAYS EVENTS, one addressing the Jewish nation, and the other addressing the world at large. But he waited until the end of his prophecy regarding the events of the end of the world before making the important statement about ***THIS GENERATION***, a clear evidence that he did not want it to be applied only to the last generation of the Jewish nation.

In Matthew's genealogy, Christ is the LATEST AND EVENTUALLY THE LAST OF THE LINEAGE AND GENERATION FROM ADAM TO ABRAHAM TO DAVID AND

TO HIMSELF(See Matthew 1). His generation was the LATEST AND WILL BE THE LAST GENERATION OF THE NATION. The significance of this genealogy is glossed over because of the general apathy towards this subject and its seemingly "boring" nature. But God inspired Matthew to record that genealogy because Jesus wanted to remind the people where in time they were living. Between Genealogy and Chronology, many important knowledge regarding the age of this world and when it will end are provided to the patient and diligent student, as this study proves. Paul stated that Christ came and died AT THE END OF THE WORLD, thereby indicating that the period of the Great Controversy and the world is divided into TWO WORLDS, the WORLD BEFORE THE OFFICIAL REVELATION OF THE KINGDOM OF GOD and the ONE POST TO IT. So, the LAST GENERATION OF THE JEWISH NATION began with the birth of Christ. Jesus's date of birth marked the beginning of the last generation of the nation. This is what he was telling them.

ELLEN G. WHITE'S 4000, 6000 YEARS PROPHECY PROVIDES 2000 YEARS LINK BETWEEN THE TWO LAST GENERATIONS AND THEIR MAJOR GREAT CONTROVERSY EVENTS

While we will not employ this prophecy to determine a doctrine, we can believe that it is God's message because her prophetic gift and its authenticity has been severally proved and verified in the church. Non-Seventh-day Adventists are at liberty to doubt and reject this fact. But when this prophecy and its veracity is proved by this study's accuracy, then these doubters will have no recourse other than to admit that Ellen G. White is indeed a true prophet. In these prophecies that God gave His servant Ellen G. White, 2000 years(between the 4000 years and 6000 years) provide the link between the two last generations and their important Great Controversy events. This is the main purpose God intended in giving her these prophecies. They were never to point to the YEAR, DAY AND HOUR OF JESUS' COMING.

The 4000, 6000 years prophecy was therefore intended to link Daniel 9 and Daniel 12, the two main prophecies that give us the TWO LAST GENERATIONS. Unfortunately, the church's historical and flawed interpretation of Daniel 12 PROPHETIC PERIODS has hampered this knowledge about God's intention for giving Ellen G. White these prophecies, ESPECIALLY THE ERROR ON WHAT THE DAILY AND ABOMINATION OF DESOLATION ARE. Based on the understanding that Jesus' Matthew 24 prophetic discourse was given to have dual application, first to the Jewish nation, and later to the world, this study is of the opinion that Jesus expects the church to also ascertain when the LAST GENERATION OF THIS WORLD began, in parallel to that of the Jewish nation. Doing this requires the detailed and accurate study of the CHRONOLOGY OF THE EARTH which has been provided in this book.

The author of this study, therefore, does not believe that the 4000, 6000 years statements on the age of the earth and the Great Controversy, are associated with, or based on, or support the "CREATION WEEK ANALOGY", which assumes that the CREATION WEEK represents SEVEN THOUSAND YEARS REDEMPTION CALENDAR. This theory supposes that each of the six creation days represent one thousand years of the Great Controversy, and the seventh day represent the millennium or one thousand years revealed to John in Revelation 20. Thus the conclusion is that Jesus's second coming will occur at exactly 6000th year of the earth. Giving that every possible calculation of the chronology of the earth, as studiously and accurately done in this study, puts the 6000th year of the earth and the Great controversy in the past(as it is commonly said, "the math is not there"), and since I believe, like other true Adventists, that the 4000, 6000 years statements by Ellen G. White are inspired and prophetic, even if they were not given to provide, ***by themselves alone***, the express or precise year of Jesus' coming, one plausible purpose for those messages must be for the church to know when the last generation of the Great Controversy begins in relation to when the last generation of the Jewish nation began. They therefore provide a 2000-year-old link between the two last generations.

In the case of the Jewish nation, the last generation began with the birth of Jesus Christ, which occurred in 2(+-1) B.C., and lasted 70 years, 10 years short of its normal length of 80 years, just as Christ prophesied. Two thousand years later brings us to AD 1999. Thus the latest generation of the earth, as at this year AD 2025, which from THIS STUDY'S accurate chronology, began AD 2000, clearly support the 2000-year link from Ellen G. White's prophecy, providing that this is the last generation of the world.

So, is it the last generation? How will we know? The answer is that the knowledge will come from Daniel 12. If the THIRD AND LAST TAKING AWAY OF THE DAILY happens in this latest generation of the earth, exactly in the year AD 2031, two thousand years since the first occurrence in AD 31, then we have our confirmation that the prophet was correct. But unfortunately, Seventh-day Adventist church's flawed interpretation of the DAILY and ABOMINATION OF DESOLATION has prevented her from seeing the connection between the 4000, 6000 years vision and Christ's crucifixion in AD 31.

Are we seeing signs that the DAILY IS ABOUT TO BE TAKEN AWAY IN THE FORM OF THE UNION OF CHURCH AND STATE TO TAKE AWAY GOD AND CHRIST(THE DAILY OF THE SANCTUARY AND HEAD OF THE CHURCH ON EARTH) AND LEGISLATION OF SUNDAY LAW AND WORSHIP? Of course the signs are everywhere and around us. The movement to unite church and state and the agitation for Sunday

Law and worship have gained popularity very rapidly. But Seventh-day Adventists are still cherishing a lot of wrong interpretations of the books of Daniel and Revelation, especially the subject of Daniel 12 prophetic periods. In another study, most of these errors have been addressed. For this, the reader is referred to the book: **Doctrinal Revelations For God's End-Time Church:** ***Exposing The Chruch's Grave Danger And Avoiding Great Disappointment 2.0.***

Thus, the 2000 years link tells us that in AD 2031, the third and last prophetic event of TAKING WAY OF THE DAILY AND SETTING UP ABOMINATION OF DESOLATION will occur. This in turn provides the link between Daniel 9 and Daniel 12 prophetic periods of the 1290, 1260, and 1335 days. With accurate arrangement of Daniel 12 prophetic days, we therefore have the last days calendar for the world beginning in AD 2031 through AD 2038, when Michael(Jesus Christ) will stand up, followed quickly with his SECOND coming, resurrection of the dead and translation of all saints to immortality.

Again, unfortunately, Seventh-day Adventist church's HISTORICAL INTERPRETATION THEOLOGY that see Daniel 12 prophetic periods as years that have already been fulfilled in the past, denies her this glorious truth. Later in this study we will provide the ARRANGEMENT OF THE DANIEL 12 PROPHETIC PERIODS. But at this juncture the reader will be provided the details of the CHRONOLOGY done in this study. It is such an amazing work and experience as the reader will see. The constructive critique of Ussher's work, revealing its flaws, while not discrediting the monumentality of that work, will be seen by the reader.

DETAILED CHRONOLOGY OF THE EARTH AND THE LAST GENERATION OF THE EARTH AND GREAT CONTROVERSY-WHY JESUS CAN NEVER COME AT THE

6000TH YEAR OF THE EARTH OR OF THE GREAT CONTROVERSY-"THE MATH IS SIMPLY NOT THERE"

EXPLAINING THE DETAILS OF THE STUDY ON CHRONOLOGY OF THE EARTH AND ITS IMPORTANT CONCLUSIONS

IMPORTANT NOTE TO THE READER: Ussher's chronology was studied, and its mistakes were corrected. This study will not offer all the details that Ussher himself has accurately and correctly covered. It will focus on the flaws of his study. Therefore this study will employ the figures and calculations that Ussher got right. But, as pointed out earlier, Ussher did make some significant mistakes that grossly affect the ultimate purpose of his work, which is the determination of the accurate age of the earth.

One of the mistakes involves the omission of about 190 years between Abraham and the beginning of the sojourn in Egypt. Isaac's age when he begat Jacob(60 years), and Jacob's age when he entered Egypt(130 years) are both missing in Ussher's chronology, particularly huge omissions and therefore big flaw in his chronology. The second mistake is the inaccurate historical date he anchored the biblical chronology to, that is, the date Jerusalem fell to Babylon and Nebuchadnezzar. Ussher chose 586 B.C. As will be seen in this study, 527 B.C. is the actual date that Jerusalem was destroyed, contrary to all the dates that have been provided by historians. This study relies on scriptures to fix when the 70 years of captivity began, instead of depending on the flawed dating system of historians such as the Canon of Ptolemy and others. A more reliable and later historic dating, those of Roman historians, complemented by accurate interpretation of Daniel 9, are seen by this author as providing more credible historic date for fixing the beginning of the 70 years of Babylonian captivity, as well as when to date Cyrus's decree. More details on these two major mistakes by Ussher will be provided further in the study.

Ussher's study is primarily on the age of the earth, and not on the age of the Great Controversy. However because he assumed that Adam sinned on the 10th day (or latest, on the first year) from creation, his chronology is also approximately that of the Great Controversy. By Great controversy is meant when Adam and Eve sinned, beginning this controversy between God and Christ on one side and Satan on the other, both parties fighting for the soul of humanity. From the chronology of the earth, with Adam's age when he sinned determined, the age of the Great controversy can easily be gotten by simply subtracting the age at which Adam sinned. In the second part of this study, it will be shown that Ussher's chronology does not give the accurate age of the Great Controversy. While Ussher determined that Adam must have sinned, at the latest, in the first year of creation, this study will show that Adam lived as long as 98 YEARS, before he sinned.

Before presenting the revised numbers and calculations that give the correct age of the earth, it will be pertinent for the reader to acquaint himself or herself with Ussher's work. His work, in its most recently revised form, is found in the book, ***ANNALS OF THE WORLD***. The reader is encouraged to read this work. It is a masterful project.

CHRONOLOGY FROM ADAM TO EXODUS

THE KEY PIECES OF THE CHRONOLOGY FROM CREATION (ADAM) TO EXODUS: ADDRESSING SEEMING DISCREPANCIES IN THE RELEVANT SCRIPTURES

The chronology from Adam to Exodus has important pieces. These pieces and the scriptures they are based on will be examined. Some of these scriptures appear to be contradictory, or appear to be contradicted by other scriptures. Let us examine these and explain the apparent contradictions (For more detailed explanations that address all the possible objections, look up Ussher's work, ANNALS OF THE EARTH. However, Ussher did make very significant errors in his study, as we shall see soon). Most of these scriptures are simple and straightforward. Therefore, we will concentrate on the seemingly challenging ones.

NOAH AND THE FLOOD

Here are the relevant scriptures.

Genesis 5:32 KJVS

"And Noah was five hundred years old: and Noah begat Shem, Ham, and Japheth".

Genesis 7:6 KJVS

"And Noah was six hundred years old when the flood of waters was upon the earth".

Genesis 7:11 KJVS

"In the six hundredth year of Noah's life, in the second month, the seventeenth day of the month, the same day were all the fountains of the great deep broken up, and the windows of heaven were opened".

Genesis 8:13-14 KJVS

"And it came to pass in the six hundredth and first year, in the first month, the first day of the month, the waters were dried up from off the earth: and Noah removed the covering of the ark, and looked, and, behold, the face of the ground was dry. [14] And in the second month, on the seven and twentieth day of the month, was the earth dried".

NOTE: The flood lasted 1 year and 10 days.

Genesis 10:21 KJVS

"Unto Shem also, the father of all the children of Eber, the brother of Japheth the elder, even to him were children born".

NOTE: Japhet is Shem's elder brother. By how many years? We will soon find out.

Genesis 11:10 KJVS

"These are the generations of Shem: Shem was an hundred years old, and begat Arphaxad two years after the flood:"

NOTE: The phrase, ***"two years after the flood"*** means two years counting from the first day of the flood, and not from the day the flood ended. How do we know this? Because Noah was 600 years old when the flood began and he was 500 years when he begat the first of his three sons, which must be Japhet because he is older than Shem. So, at the beginning of the flood, Japhet was 100 years old. The flood lasted 1 year plus 10 days. Add another year and you will have Japhet being 102 years old "two years after the flood". With Shem being 100 years old "two years after the flood", it is easy to see that the phrase means two years from the first day of the flood. So, Japhet was two years older than Shem.

TWO WAYS TO BRIDGE NOAH AND ARPHAXAD

1. We can simply connect Noah, The flood and ARPHAXAD by 600 + 2 + 35(Arphaxad's age when he begat Salah). What this does is that Noah's age before the Flood encompasses Shem's age at the beginning of the flood. Since ARPHAXAD is born 2 years after the flood, there is no need to include Shem's age in the chronology.

2. Alternatively, we can include Shem's age at the beginning of the flood, which will be 98 years. So, we will have: 502(Noah's age when he begat Shem) + 100(Shem's age when he begat ARPHAXAD. In this case, we don't need to include the 2 years because it is included in the 100 years.

Either way, the same result is achieved.

FROM ABRAHAM TO EXODUS

Under this section, we will discover why and how Ussher and other chronologists made significant mistakes, omitting many years in the chronology.

USSHER'S MISTAKE

Ussher in his study assert that the 430 years sojourn began when Abraham, at 75 years, departed from Haran. Apparently Ussher considered that Abraham's journey into Egypt, soon after departing from Haran, qualifies to be included in the 430 years sojourn God had declared to him. But, this view by Ussher is so strangely and utterly far from the truth on this matter. It must be that Ussher simply did not do enough searching and research, or simply could not reconcile Paul's narrative on the subject (See Galatians 3:15-17). There is an unbroken chronology between Abraham, Isaac, Jacob, and the beginning of the 430 years sojourn in Egypt. Therefore Paul's statement needs to harmonize with Genesis narrative because the Genesis narrative is the first narrative. Paul, on the other hand, is not focusing on accuracy of chronology, as the Genesis narrative is.

Scriptures make it absolutely clear that neither Abraham, nor Isaac after him, ***SOJOURNED*** (made a home in and lived out their lives) in Egypt. Abraham's journey into Egypt was very brief (which therefore does not qualify as **sojourning**), and Isaac himself was warned by God, never to set foot in Egypt, and he never did. It therefore makes no sense whatsoever that the 430 years should be counted from Abraham. In God's declaration, there is no sense that He meant anything but a ***continual stay*** in Egypt until the exodus. This is what qualifies a chronology to fulfill the prophecy of the 430 years. But the fact that Isaac never set foot in Egypt disrupts any such chronology. Sojourning in this context means, a continual stay and living out one's life in a place. Therefore, Abraham's very brief journey into Egypt, and **ISAAC'S ZERO ENTRANCE INTO EGYPT**, indicate that the fulfillment of the 430 years prophecy never began both of them.

Instead, here are the various scriptures that show unbroken chronology from Abraham to when the 430 years actually began:

ABRAHAM'S BRIEF SOJOURN IN EGYPT - Genesis 12:10 KJVS

"And there was a famine in the land: and Abram went down into Egypt to sojourn there; for the famine was grievous in the land" (Read through rest of the chapter about what happened to Abraham in Egypt).

Then in chapter 13:1, we see that Abraham did not remain in Egypt, and when he left Egypt he was with Lot:

Genesis 13:1-2 KJVS

"And Abram went up out of Egypt, he, and his wife, and all that he had, and Lot with him, into the south. [2] And Abram was very rich in cattle, in silver, and in gold".

GOD WARNS ISAAC NOT TO SET FOOT IN EGYPT:

Genesis 26:1-6 KJVS

"And there was a famine in the land, beside the first famine that was in the days of Abraham. And Isaac went unto Abimelech king of the Philistines unto Gerar. [2] ***AND THE LORD APPEARED UNTO HIM, and said, GO NOT DOWN INTO EGYPT***; dwell in the land which I shall tell thee of: [3] Sojourn in this land, and I will be with thee, and will bless thee; for unto thee, and unto thy seed, I will give all these countries, and I will perform the oath which I sware unto Abraham thy father; [4] And I will make thy seed to multiply as the stars of heaven, and will give unto thy seed all these countries; and in thy seed shall all the nations of the earth be blessed;

[5] Because that Abraham obeyed my voice, and kept my charge, my commandments, my statutes, and my laws. [6] And Isaac dwelt in Gerar:"

The PROPHECY ABOUT THE 430 YEARS IN EGYPT WAS ABOUT ABRAHAM'S SEED, AND NOT ABRAHAM HIMSELF:

The prophecy expressly says that Abraham's seed, not Abraham himself, will sojourn and dwell in Egypt for 430 years. Here is the prophecy:

Genesis 15:13-16 KJVS

"And he said unto Abram, Know of a surety that <u>THY SEED</u> shall be a stranger in <u>A LAND</u>(not LANDS) that is not theirs, and SHALL SERVE THEM(neither Abraham nor Isaac was a slave in any nation or to any king); and they shall afflict them FOUR HUNDRED YEARS; [14] And also **<u>THAT NATION</u>** (not NATIONS), whom they shall serve, will I judge: and afterward shall they come out with great substance. [15] And thou shalt go to thy fathers in peace; thou shalt be buried in a good old age. [16] BUT IN THE FOURTH GENERATION they shall come hither again: for the iniquity of the Amorites is not yet full".

The prophecy is clear; Abraham was to die before the prophecy is fulfilled. We have already shown that Isaac never set foot in Egypt. That leaves Jacob as the person to fulfill the prophecy. Did Jacob (renamed Israel) and his children (the Children of Israel) enter into Egypt to sojourn there and multiply there, in fulfillment of the prophecy? It's a definite yes.

Here is scripture that plainly state that it was not Abraham, or even Isaac that sojourned in Egypt for 430 years:

Exodus 12:40-41 KJVS

"Now the sojourning of <u>THE CHILDREN OF ISRAEL, WHO DWELT IN EGYPT</u>, was four hundred and thirty years. [41] And it came to pass at the end of the four hundred and thirty years, even the selfsame day it came to pass, that all the hosts of the LORD went out from <u>THE LAND OF EGYPT</u>".

Abraham and Isaac cannot be described as the CHILDREN OF ISRAEL. Neither did they ***DWELL*** in Egypt, as Abraham was there only very briefly, and Isaac, not at all. The name "Israel" is Jacob's name, and the name therefore never came into existence and use until Jacob's encounter with God. Jacob is the ***DIRECT FATHER*** of the nation of Israelites. There was no nation yet until Jacob had children. Abraham was not a notion by himself, and neither was Isaac. The fulfillment of the promise of a nation-multiplication-came with Jacob.

This scripture is as clear as day, that it was not Abraham or even Isaac, that began the 430 years period. Based on Ussher's assumption that the 430 years began with Abraham, in his 75th year, it would mean that the actual AFFLICTION IN EGYPT was for just 215 years, that is 430 - (25 + 60 + 130)(Abraham was 100 years old when he begat Isaac, and Isaac was 60 years old when he begat Jacob, and Jacob was 130 years when he and his children and household of 75 people entered Egypt to sojourn there, with Joseph already there). This clearly contradicts God's declaration to Abraham that his seed will be afflicted for(approximately) 400 years. By rule of approximation, 215 is in no way approximately 400 years, not even close.

Granted it was not until Joseph and his brothers had all died that an evil Pharaoh arose and the affliction began. With Joseph being in his late thirties when Jacob and his brethren came to live in Egypt with him, and being that Joseph died at the age of one hundred and ten years, the actual AFFLICTION of the people was actually (430 - 70 = 360 years). This, according to the rule of approximation, is 400 years or four generations. The 60 years part falls into the fourth generation because it is greater than 50. This is more accurate than the implication from Ussher's view, which is that the AFFLICTION part of the sojourn was only for 215 years, which will be approximately 300 years or three generations, in clear contradiction of God's prophecy to Abraham.

WHAT ABOUT ACTS 7:4 AND GALATIANS 3:17?

These are the texts that misled Ussher and other chronologists because they did not compare them with the old testament scriptures or even with the immediate context. As a rule, scripture should interprets scripture.

Galatians 3:15-17 KJVS

"Brethren, I speak after the manner of men; Though it be but a man's covenant, yet if it be confirmed, no man disannulleth, or addeth thereto. [16] Now to Abraham and his seed were the promises made. He saith not, And to seeds, as of many; but as of one, And to thy seed, which is Christ. [17] And this I say, that the covenant, that was confirmed before of God **IN CHRIST**, the law, which was four hundred and thirty years after, cannot disannul, that it should make the promise of none effect".

Some have used the above scriptures to justify counting the 430 years from when God RE-AFFIRMED the everlasting covenant to Abraham, which is at his 99TH YEAR (Genesis 17). But what these teachers miss is that Paul was alluding to the fact that this covenant has been REAFFIRMED to every (chosen "SEED") descendants of Adam, starting with SHEM.

Notice that Paul did not say, "CONFIRMED BEFORE OF GOD IN ABRAHAM", because Abraham is simply one of the LINE AND SUCCESSION OF "SEEDS", in whom the covenant was REAFFIRMED, temporary custodians of the everlasting covenant. Jesus is the PRIMARY CUSTODIAN of the everlasting covenant. The last SEED, to whom the covenant was reaffirmed before the sojourn in Egypt began, is Jacob. It is from this point that Paul based his statement on the 430 years. Because Paul was not focused on the chronology, he did not see the need to provide this explanation.

So, Paul simply was saying that this covenant was reaffirmed just before the 430 years sojourn began. Sure enough, we see that just before Jacob entered Egypt with his children to fulfill the 430 years prophecy, God reaffirmed the covenant of promise with him.

We read this as follows:

Genesis 46:2-4 KJVS

"And God spake unto ISRAEL in the visions of the night, and said, Jacob, Jacob. And he said, Here am I. [3] And he said, I am God, the God of thy father: fear not to go down into Egypt; for I WILL THERE MAKE OF THEE A GREAT NATION: [4] <u>I WILL GO DOWN WITH THEE INTO EGYPT</u>; and <u>I WILL ALSO SURELY BRING THEE UP AGAIN</u>: and Joseph shall a put his hand upon thine eyes".

In harmony with what God had declared to Abraham, we see how beautifully this prophecy that God gave to Abraham, is fulfilled by Jacob, the earthly seed of Abraham, and by Jesus, the CHIEF(SPIRITUAL)SEED OF ABRAHAM. It was Jesus in his pre-human days who spoke to Jacob. So, both the natural seed of Abraham, Jacob, as well as the spiritual seed of Abraham, Jesus, entered Egypt, in fulfillment of the prophecy, with Jesus entering with Jacob and his children spiritually and invisibly. This is what Paul was referring to, and not that Abraham's 99th year marked the beginning of the 430 years.

NO CONTRADICTION BETWEEN STEPHEN'S ACCOUNT AND GENESIS ACCOUNT OF WHEN ABRAHAM LEFT HARAN: There is no contradiction between Stephen and Genesis account of when Abraham left Haran.

"Then said the high priest, Are these things so? [2] And he said, Men, brethren, and fathers, hearken; The God of glory appeared unto our father Abraham, when he was in Mesopotamia, before he dwelt in Charran, [3] And said unto him, Get thee out of thy country, and from thy kindred, and come into the land which I shall shew thee. [4] Then came he out of the land of the Chaldaeans, and dwelt in Charran: and from thence, when his father was dead, he removed him into this land, wherein ye now dwell. [5] And he gave him none inheritance in it, no, not so much as to set his foot on: yet he promised that he would give it to him for a possession, and to his seed after him, when as yet he had no child. [6] **And God spake on this wise, That HIS SEED should sojourn in A STRANGE LAND; and that they should bring them into bondage, and entreat them evil four hundred years. [7] And THE NATION to whom they shall be in bondage will I judge, said God: and after that shall they come forth, and serve me in this place. [8] And he gave him the covenant of circumcision: AND SO ABRAHAM BEGAT ISAAC, AND CIRCUMCISED HIM THE EIGHTH DAY; AND ISAAC BEGAT JACOB; AND JACOB BEGAT THE TWELVE PATRIARCHS".**

Ussher got it right with his explanation that Abraham was the oldest child of Terah. Terah simply had Abraham when he was 130 years old. The reason Abraham is listed first is not that he is the first born, but rather it is customary with chronology to list the person through whom the particular lineage is perpetrated, in this case the lineage that extended to Jesus Christ(See Matthew 1)

THE UNBROKEN CHRONOLOGY FROM ABRAHAM TO WHEN THE 430 YEARS BEGAN (WHEN JACOB ENTERED EGYPT TO DWELL THERE)

Genesis 21:5 KJVS

"And Abraham was **AN HUNDRED YEARS OLD**, when his son Isaac was born unto him".

Genesis 25:24-26 KJVS

"And when her days to be delivered were fulfilled, behold, there were twins in her womb. [25] And the first came out red, all over like an hairy garment; and they called his name Esau. [26] And after that came his brother out, and his hand took hold on Esau's heel; and his name was called Jacob: and **ISAAC WAS THREE SCORE YEARS OLD WHEN SHE(REBECCA) BARE THEM (ESAU AND JACOB)".**

Genesis 47:7-10 KJVS

"And Joseph brought in Jacob his father, and set him before Pharaoh: and Jacob blessed Pharaoh. [8] And Pharaoh said unto Jacob, How old art thou? [9] And Jacob said unto Pharaoh, The days of the years of my pilgrimage are **AN HUNDRED AND THIRTY YEARS**: few and evil have the days of the years of my life been, and have not attained unto the days of the years of the life of my fathers in the days of their pilgrimage. [10] And Jacob blessed Pharaoh, and went out from before Pharaoh".

So, from Abraham we have the following chronology, up to the Exodus:

100 + 60 + 130 + 430

This is the part of chronology that Ussher (and other chronologists) missed, and which significantly flaws his work. The whole chronology from Adam to Exodus is beautifully depicted in a chart and graph at the end of the book.

THE FOURTH GENERATION-FROM WHAT POINT OF RECKONING?

The whole sojourn in Egypt was exactly 430 years, and it started with the entrance of Jacob and his sons after the latter learnt that Joseph was still alive (See Genesis 46:1-4; Exodus 1:1-8; 12: 40,41). But God's promise to Abraham states that the afflictions will be for "four hundred years", and in fact it is in the four hundredth year that they will be delivered, meaning that their afflictions will not be a complete four hundred years. Anywhere after 300 years will be considered the four hundredth year or fourth generation. If the whole 430 years are taken as how long they were enslaved, this will contradict God's prophecy, because anywhere after 400 years is considered the 500th year or the fifth generation.

SUMMARY OF THE CHRONOLOGY FROM ADAM TO EXODUS: 2728 YEARS.

BELOW IS THE PICTOGRAM

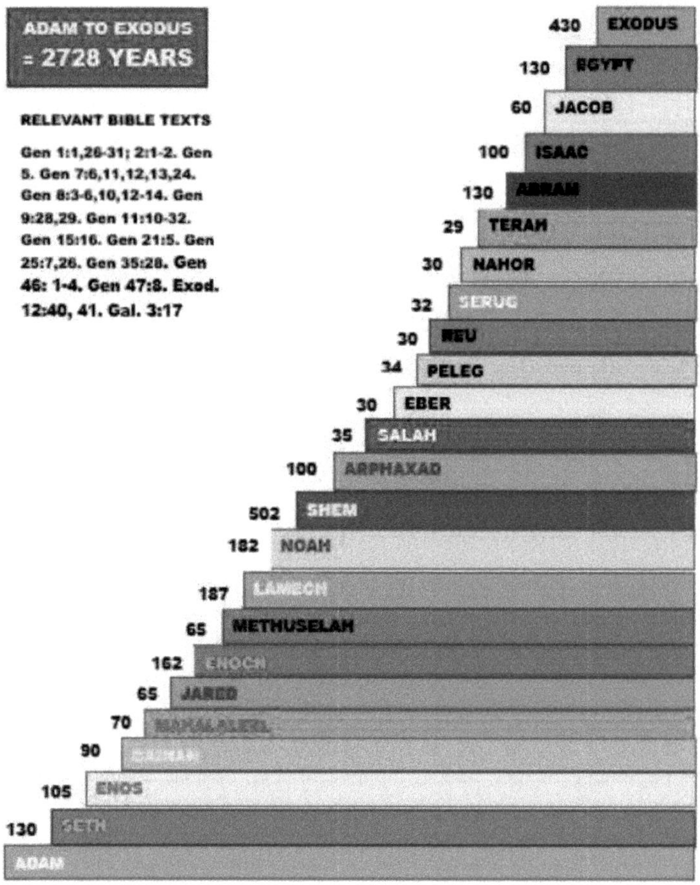

SECTION 2
OF THE CHRONOLOGY OF THE EARTH

THE DIVIDED KINGDOM AND HOW LONG IT LASTED

Here, Ussher was correct in choosing Ezekiel's account, for no one could know better than Ezekiel because he had access to the chronicles of the kings. So, let us read what Ezekiel wrote, which is actually what God himself told him, about how long the divided kingdoms lasted.

Ezekiel 4:1-17 KJVS

"Thou also, son of man, take thee a tile, and lay it before thee, and pourtray upon it the city, even Jerusalem: [2] And lay siege against it, and build a fort against it, and cast a mount against it; set the camp also against it, and set battering rams against it round about. [3] Moreover take thou unto thee an iron pan, and set it for a wall of iron between thee and the city: and set thy face against it, and it shall be besieged, and thou shalt lay siege against it. This shall be a sign to the house of Israel. [4] Lie thou also upon thy left side, and lay **THE INIQUITY OF THE HOUSE OF ISRAEL** upon it: according to the number of the days that thou shalt lie upon it thou shalt bear their iniquity. [5] For I have laid upon thee **THE YEARS OF THEIR INIQUITY**, according to the number of the days, **THREE HUNDRED AND NINETY DAYS**: so shalt thou bear the iniquity of the house of Israel. [6] And when thou hast accomplished them, lie again on thy right side, and thou shalt bear the **INIQUITY OF THE HOUSE OF JUDAH FORTY DAYS**: **I HAVE APPOINTED THEE EACH DAY FOR A YEAR".**

When the reigning years of the kings, from Rehoboam to Zedekiah's 11th year, are tallied, one will get 393 years, without factoring in the overlaps due to viceroy years (years that a prince began to reign as king while the father was still alive but incapable of accomplishing the royal task). When those overlaps are factored in, it will be seen that the chronology of the divided kingdom harmonizes with God and Ezekiel's 390 years. They should know better any how.

SECTION THREE OF THE CHRONOLOGY

WHEN TO BEGIN COUNTING THE 70 YEARS OF BABYLONIAN CAPTIVITY

Some presume that it is from the day of first invasion(siege) when Daniel and other princes of Israel were taken to Babylon. Others decide to choose the end of Jehoiakim's 11 years, when he was put in shackles and taken to Babylon. In this study we will show that the seventy years actually began on the last day of the second siege, which is at the end of Mattaniah (Zedekiah's) 11 years reign.

Daniel wrote:

Daniel 9:2 KJVS

"In the first year of his reign I Daniel understood by book the number of the years, whereof the word of the LORD came to Jeremiah the prophet, that he would accomplish seventy years
IN THE DESOLATIONS OF JERUSALEM".

There, in plain language, we have that the 70 years began with the destruction of the city and the temple. This did not occur with the first invasion and siege.

Concerning the first invasion by Nebuchadnezzar, here is what Daniel narrated, which exactly corroborated the book of the Chronicles:

Daniel 1:1-2 KJVS

"In the third year of the reign of Jehoiakim king of Judah came Nebuchadnezzar king of Babylon unto Jerusalem, and besieged it. [2] And the Lord gave Jehoiakim king of Judah into his hand, with part of the vessels of the house of God: which he carried into the land of Shinar to the house of his god; and he brought the vessels into the treasure house of his god".

2 Chronicles 36:5-7 KJVS

"Jehoiakim was twenty and five years old when he began to reign, **and he reigned eleven years in Jerusalem**: and he did that which was evil in the sight of the LORD his God. [6] Against him came up Nebuchadnezzar king of Babylon, and bound him in fetters, to carry him to Babylon. [7] **Nebuchadnezzar also** carried of the vessels of the house of the LORD to Babylon, and put them in his temple at Babylon".

A NOTE TO EXPLAIN THE APPARENT(SEEMING) DISCREPANCY OF DANIEL'S ACCOUNT WITH THAT IN 2 CHRONICLES

As Daniel wrote, Nebuchadnezzar invaded Israel in Jehoiakim's third year. But Zedekiah did reign for 11 years. This would not be the case if Nebuchadnezzar had put him in fetters in third year of his rule. But Nebuchadnezzar did not put him in shackles and take him to Babylon in that third year of his rule. He permitted Jehoiakim to continue to rule on the condition that he would not rebel against Babylonian authority. Compliance will of course involve paying tributes and taxes to Babylon. This is the same situation the nation was under Rome. The government was still functioning, but subject to the grace of Nebuchadnezzar. Zedekiah therefore continued to rule for another eight years. But in his eleventh year, Jehoiakim rebelled, and that is when Nebuchadnezzar deposed him, had him in fetters, took him to Babylon, and put his son Jehoiachin as king. It is also at that time that Nebuchadnezzar began to confiscate the vessels of the temple, and took the princes of Israel, Daniel, and the others to Babylon, as early as Jehoiakim's third year, which is also when he took Daniel and the others to Babylon.

DISCREPANCY ON JEHOIACHIN'S AGE WHEN HE BEGAN TO RULE

While 2 Chronicles have it as 8 years, 2 Kings has it as 18 years. This is clearly a typographical error. What makes logical sense is that the correct age is 18 years. The typo error was made in the 2 Chronicles account.

Zedekiah, JEHOIACHIN'S brother was 21 years when he began to rule, and 18 years seem to be the age that makes more sense, or else it would mean that Zedekiah was older than JEHOIACHIN by as much as 13 years. What makes more sense instead is that the age difference is just 3 years.

Here is another clear evidence that the counting of the 70 years began with the destruction of Jerusalem and the temple in the last year of Zedekiah's 11-year reign:

2 Chronicles 36:19-21 KJVS

"And they burnt the house of God, and brake down the wall of Jerusalem, and burnt all the palaces thereof with fire, and destroyed all the goodly vessels thereof. [20] And them that had escaped from the sword carried he away to Babylon; where they were servants to him and his sons until the reign of the kingdom of Persia: [21] **To fulfil the word of the LORD by the mouth of Jeremiah, until the land had enjoyed her sabbaths: for as long as she lay desolate she kept sabbath, to fulfil threescore and ten years".**

Note the statement, "for as long as she (the land) lay desolate, she kept sabbath". The land cannot lie desolate while there are inhabitants in it, and is being cultivated, and while a functional government still existed. It was not until the end of Zedekiah's 11 years that the land became desolate. This is when the land became empty, as is evident from the statement, “And them that had escaped from the sword carried he away to Babylon”. At this point, no one was left in Israel. All government, all society and life ended in Israel. It makes perfect sense, therefore, that this is when to begin counting the 70 years of captivity. Note that the reason the land was not keeping her sabbath was because of the evil rules of evil kings, of which Zedekiah was the last one. So the land could not begin keeping her sabbath until Zedekiah’s rule ended.

Now that we know that the seventy years began at the end of Zedekiah’s reign, which is 19 years after the first siege, it is therefore clear that even if we employ the dates of historians, the correct date for the beginning of the 70 years would be 586/587 B.C., as Ussher determined. However neither 606 B.C., 596 B.C., nor 586 B.C., is the accurate year the 70 years began, because historians are wrong. Their dating do not harmonize with Daniel 9:25, Isaiah and Jeremiah's prophecy. More will be said about this later.

The year, 527 B.C. is the year chosen by this author because it harmonizes with Danie, Isaiah, l and Jeremiah's prophecies that Cyrus’ decree is the decree that fulfills Daniel 9:25 and therefore is the decree that bears the date of 457 B.C., and not Artaxerxes’ decree. Archeology, on which these dates and dating are based, is a very faulty and inaccurate science. God couldn’t possibly allow this important event and date to be determined by Archeology. This why God gave us surer dates such as 27 A.D., and 31 A.D., which are the dates of Jesus’ baptism, and death, which we do not need such uncertain events such as the counting of eclipses to determine. These dates are accurately documented by historians of the Roman period, which are not only more recent but have also been better preserved. Going back from these dates William Miller arrived at the 457 B.C. date, and with the prophecies of Daniel and Jeremiah on the 70 years, we accurately place the beginning of the 70 years at 527 B.C. (457 + 70). More will be said about this further in this study. Thus we do not need the inaccurate dating by Ptolemy.

SECTION 4

OF THE CHRONOLOGY OF THE EARTH

THE GREGORIAN TERMINAL DATE FOR THE BIBLICAL CHRONOLOGY: WHERE BIBLICAL CHRONOLOGY LINKS UP WITH SECULAR HISTORICAL DATE

THE IMPORTANCE OF WILLIAM MILLER'S 2300-DAY PROPHECY STUDY-THE NON-RELIABILITY OF HISTORICAL DATES

It will be expected that God, having meticulously given biblical chronology from Adam, will no less ensure that the historical date upon which it will be linked to modern date, has to be a reliable one. Historians have given us many and varying dates for when Jerusalem fell to Nebuchadnezzar. This event happens to be the most important date on which to link biblical part of the chronology of the earth. Some of these dates(years) are as follows: 609 B.C., 606 B.C., 588 B.C., 586 B.C., etc. Ussher happens to have chosen 586 B.C. If the Devil is to ensure that the computation of the chronology of the earth will not be accurately obtained, one of the important places he will misrepresent to theologians and historians is the event and or date the biblical chronology is joined to historical dating. This explains why this important event is given various and varying dates. But as should be expected, God ensured that this obstacle will not stand in the way of the truth on this matter.

Not surprisingly, Satan has ensured that historians will muddy up this important date. Satan has reason to do this and that is that he does not want the actual age of the earth to be known. He is the architect of the flawed carbon dating system by which archaeologists and other scientists base their dating of the earth and the universe as being in the MILLION to BILLIONS OF YEARS. These dating methods are so very imperfect. But Seventh-day Adventists church, and any other group that values the scriptures, is in a unique and vantage point that helps her determine the correct dates at which these events actually occurred. That vantage point is provided by a combination of William Miller's work on the 2300-DAYS PROPHECY of Daniel 8:13,14, and the Spirit of Prophecy, given to the church through Ellen G. White. The key is to work backwards from dates that are more reliable and recent and do not need the inaccurate science of astronomy and carbon dating, and that do not need the counting of eclipses. Those dates are those of Roman historians or Historians of the Roman empire era, an era closest to Jesus's life and ministry on earth.

God knows that historical dating will be important to complete the dating of the earth. This is why God has ensured that some historical dates and dating that will be needed for the chronological work are accurate, especially the dating of the BIRTH, BAPTISM, and DEATH of Jesus Christ. Likewise the dates for the first and second sieges of Jerusalem by the Roman army in 66 and 70 C.E respectively are accurately recorded, and these dates are very recent, unlike the Ptolemy canon dating. God has ensured that these dates were accurately chronicled because they are needed for the interpretation of the prophecies of Daniel 8-12.

Based on William Miller's work, the start date for the 2300 prophetic days (2300 YEARS), is 457 B.C. The details of that study will not be undertaken here as the reader can read it elsewhere. But the important aspects, as well as where that study has minor errors, will be mentioned. William Miller's choice of 457 B.C. is based on when **Daniel 9:25** was fulfilled. In other words, it is the date when the decree to build and restore Jerusalem, after the Babylonian captivity, was fulfilled. From this date, the last seven years of the first 490 years of the 2300 years, appointed for Christ to confirm the new covenant, began in 27 A.D., and ended in 34 A.D. But as the prophecy states, in the middle of this seven-year period, Jesus was to be crucified, which was in 31 A.D. God has preserved these dates as carefully as He preserved the scriptures.

After Jesus was crucified, resurrected, and glorified, he completed the remaining 3.5 years of the seven years through empowering his disciples. So, he did the first part of the seven-year ministry in person, and the second part he accomplished as Divinity, working through his disciples by indwelling and guiding them as the omnipresent person, the Holy Spirit. Thus, we have those THREE IMPORTANT DATES AND EVENTS OF THE GREAT CONTROVERSY, namely, **27 A.D., 31 A.D., and 34 A.D.,** locked in. No other church is blessed with these dates, as much as Seventh-day Adventist church, and no other church believe in these dates as Seventh-day Adventist church does. Seventh-day Adventist church believe strongly in these dates, because as Daniel's prophecy states, they the dates and the events Daniel 8 vision which seals the vision, meaning making it certain, assured and unimpeachable. Here is how Daniel's prophecy puts it:

"Seventy weeks are determined upon thy people and upon thy holy city, to finish the transgression, and to make an ed of sins, and to make reconciliation for iniquity, and to bring in everlasting righteousness, ***AND TO SEAL UP THE VISION***(of the evening and morning) ***AND PROPHECY***. And to anoint the most holy" – Daniel 9:24. Emphasis mine.

The implication of this is very huge. It means that the events of this 70 weeks of years constitute the heart and key of the entire vision, meaning that when their events comes to pass and are well documented and verified, they make sure and certain the fulfillment of the rest of the vision. It is no wonder then that God in His infinite wisdom ensured that these events and their dates are not involved in any speculation or uncertainty. In his book Daniel and Revelation, Uriah Smith documents how historians of the Roman era, that lived very close, and some in the very days of Christ's fulfillment of the last week of the 70 weeks, accurately chronicled and dated those events of the last seven years of Christ's ministry in fulfillment of Daniel 9. In his well-articulated writing on this, here is Uriah Smith as he quoted Dr Haly,

"It is not to be questioned that our Saviour attended every Passover that occurred during his public ministry; and we have mention of only four such occasions previous to his crucifixion. These are found in the following passages: John2:13; 5:1; 6:4; 13:1. At the last-mentioned Passover he was crucified. From facts already established, let us then see where this would locate the crucifixion. As he began his ministry in the autumn of AD 27, his first Passover would occur the following spring, AD 28; his second, AD 29, his third AD 30; and his fourth and last, AD 31. This gives us three years and a half for his public ministry, and corresponds exactly to the prophecy that he should be cut off in the midst, or middle, of the seventieth week. As the week of years commenced in the autumn of AD 27, the middle of the week would occur three and one half years later, in the spring of 31, where the crucifixion took place. Dr. Hales quotes Eusebius, A.D.300, as saying: "It is recorded in history that the whole time of our Saviour's teaching and working miracles was three years and a half, which is the half of a week [of years]. This, John the evangelist will represent to those who critically attend to his Gospel." Of the unnatural darkness which occurred at the crucifixion, Hales, Vol.I, pp.69,70, thus speaks: "Hence it appears that the darkness which 'overspread the whole land of Judea' at the time of our Lord's crucifixion was preternatural, 'from the sixth until the ninth hour,' or from noon till three in the afternoon, in its duration, and also in its time, about full moon, when the moon could not possibly eclipse the sun. The time it happened, and the fact itself, are recorded in a curious and valuable passage of a respectable Roman Consul, Aurelius Cassiodorius Senator, about A.D.514: 'In the consulate of Tiberius Caesar Aug. V and AElius Sejanus (U.C.784, A.D.31), our Lord Jesus Christ suffered, on the 8th of the calends of April (25th March), when there happened such an eclipse of the sun as was never before nor since.' "In this year, and in this day, agree also the Council of Cesarea, A.D.196 or 198, the Alexandrian Chronicle, Maximus Monachus, Nicephorus Constantinus, Cedrenus; and in this year, but on different days, concur Eusebius and Epiphanius, followed by Kepler, Bucher, Patinus, and Petavius, some reckoning it the 10th of the calends of April, others the 13th." (See on chapter 11:22.) Here, then, are thirteen credible authorities locating the crucifixion of Christ in the spring of A.D.31. We may therefore [230] set this down as a fixed date, as the most cautious or the most skeptical could require nothing more conclusive. This being in the middle of the last week, we have simply to reckon backward three and a half years to find where sixty-nine of the weeks ended, and forward from that point three and a half years, we find ourselves in the autumn of A.D.27, where, as we have seen, the sixty-nine weeks ended, and Christ commenced his public ministry. And going from the crucifixion forward three and a half years, we are brought to the autumn of A.D.34, as the grand terminating point of the whole period of the seventy weeks. This date is marked by the martyrdom of Stephen, the formal rejection of the gospel of Christ by the Jewish Sanhedrin in the persecution of his disciples, and the turning of the apostles to the Gentiles. And these are just the events which one would expect to take place when that specified period which was cut off for the Jews, and allotted to them as a peculiar people, should fully expire"- Daniel and revelation, Uriah Smith, p 266-269.

Here we can see that the historical records of Christ's crucifixion and the dating as AD 31, is very authentic and unquestionably accurate. This accurate history along with accurate interpretation of Daniel 9 was meant by God to be enough to decode the Daniel 14 "twenty-three hundred days" prophetic period. The Canon of Ptolemy, which depends on counting of eclipses, a task that is easily involved in uncertainty from human deficiency, is not needed at all to determine when the 2300 days period began. All that is needed, as Uriah Smith indicates above, is simply to,

1. Know that the 70 weeks of Daniel 9 is cut off, that is the first subsection of the 2300 days of Daniel 8.
2. Know that the last week of the 70 weeks was Christ's last seven years to confirm the covenant, and that he died in the middle of it.
3. From that verified and well documented date and event, AD 31, we simply count backwards.

Counting backwards from AD 31, we 487 years to the date 457 BC, correcting for the non-existent 0 BC/AD. Thus there is simply no reason whatsoever to involve the Canon of Ptolemy in this calculation of the prophecy. Again counting backwards from AD 31, we have the first three and half years of the last seven years as beginning AD 27, which is when again reliable Roman history and biblical record itself indicate that our Savior was baptized at river Jordan(See Mark 1:15; Luke 3:1-23). Counting three and half years forward from this authenticated date of AD 31, we come to the last date of the last week of the 70 weeks of years, AD 34, the date that marks the end of the period given for the nation to fill the cup of her iniquity, just as the race and nation of Canaan before her did. On that date, they stoned Stephen because he presented to them a risen and gloried Christ, that Holy Spirit personality("Now the Lord is that Spirit", wrote Paul). When they rejected and crucified the human Christ, he asked God to forgive the nation because as he stated, "they know not what they do". But when they rejected and killed the Spirit-Christ in the person of Stephen, the nation filled the cup of her iniquity. Probation ended for the apostate nation and church. Subsequently she was destroyed by Rome by the permission of God, as her judgment. When they rejected and crucified Christ, they fulfilled the first of the three prophetic event of TAKING AWAY OF THE DAILY AND SETTING UP ABOMINATION OF DESOLATION, Rome, Caesar and sun god of Rome. Jesus and his father, constitute THE DAILY IN THE SANCTUARY, the ones to which all their sanctuary services represented, consisted, and held up. When the nation TOOK THEM AWAY, and declared, "We have no King but Caesar", thereby setting up the Abomination of desolation as their King, they committed the TRANSGRESSION OF DESOLATION that Daniel prophesied about in Daniel 8:10-13. Two more occurrences and fulfillment of the TAKING AWAY OF THE DAILY was to be fulfilled in the future, the third and last being the one that will fulfill soon in the CHURCH AND STATE UNION that will again acknowledge roman Caesar in sacerdotal cloth(Pope of Rome) as the head of the church and the legislation of universal Sunday law to the honor of the Sun god of Rome.

So, arriving at 457 BC start date of Daniel 8:14 prophetic period of 2300 days(years), did not need the Canon of Ptolemy at all. God had made sure that would not be necessary. The more authentic and recent Roman history was what God approved for the interpretation and computation of the prophetic period. This was designed by God to also aid the church in determining when the Babylonian captivity occurred, without depending on the uncertainty of human dating system. What the church simply needed is to understand that in Daniel 9: 25, the commandment to build and restore Jerusalem will be given by Cyrus at exactly at the end of the 70 years of Babylonian captivity(See Isaiah). This being the case, the church simply needs to add 70 years to 457 BC to get the date that Babylonian captivity began, which is 527 BC, the important date on which the biblical chronology hooks up to secular dating.

But instead of doing this, William Miller and the Millerite theologians allowed themselves to be influenced and disconcerted by the erroneous dates and dating of Archeologists and Historians. After determining that the 2300 days of Daniel 8:14 began on 457

BC, William Miller consequently ran into the problem of determining whose decree marked the beginning of the 70 weeks. Since the erroneous human dating has Cyrus decree as going out in 586 BC(some records say 588 BC, 609 BC, etc), and that of Artaxerxes in 457 BC, William Miller decided, against scriptural evidence provided by Isaiah, Jeremiah, and Daniel, that it is Artaxerxes decree that marked the beginning of the 2300 days and 70 weeks. As much as this did not disrupt the start date of 457 BC, it however disrupts biblical chronology, by assigning Cyrus decree and the destruction of Jerusalem a much earlier dates than should be the case. Since the event of the destruction of Jerusalem and beginning of the 70-year captivity constitute the link with biblical chronology, error is thereby introduced in biblical chronology studies. This is evident in Ussher's work, where he employed 586 BC as when the 70-year captivity began instead of 527 BC. A non-existent gap of years is thus introduced between when the 70 years ended and when the decree to build and restore Jerusalem unto the Messiah(Daniel 9:25) was fulfilled.

This is why, in this chronological study, this is identified as one of the errors of Ussher's work, and the correct date of 527 BC is instead employed. The proof of the correctness of 527 BC date can easily be verified in that the chronology in this book harmonizes with the scriptural note of Christ being born at the beginning of the latest and last century of the Jewish nation, which is the 53rd century from Adam. This chronology will once again be confirmed as accurate when the lates generation of the world, the 78th generation from Adam will not pass before Christ's come. The indication will come when in AD 2031, CHURCH AND STATE UNION will fulfill Daniel 12 prophetic period event of TAKING AWAY THE DAILY, thus starting the last seven years of this world's history(1290 + 1260 = 2550 DAYS or 7 YEARS).

THE 457 B.C. AND ITS RELATION TO WHEN JERUSALEM FELL TO NEBUCHADNEZZAR AND WHY IT IS CRUCIAL TO THE CHRONOLOGY OF EARTH AND GREAT CONTROVERSY

The relationship between the date 457 B.C. (when the decree was given to rebuild and restore Jerusalem after the Babylonian captivity), and when the captivity began, is 70 YEARS. This is solidly based on Isaiah and Jeremiah's visions and prophecies that the captivity was to last exactly 70 YEARS, with the end of this period marked by Cyrus's decree(not Artaxerxes's) for the rebuilding of the city and temple. Thus, 70 years of captivity has to begin in the year 527 B.C., irrespective of what historians have presented. The historians did not and still do not believe that the scriptures are accurate. But that is their choice. The church, however, does not have to go along with them. Any dates that do not align with scriptures should be discarded by the church. So based on Daniel 9:25, there should be only 70 years between when the decree to rebuild the temple (and Jerusalem) was given, and when they were destroyed by Nebuchadnezzar.

This means that Jerusalem fell to Nebuchadnezzar in 527 B.C., leaving no gap between when this event happened and when Daniel 9:25 was fulfilled.

SO WHY DO WE HAVE DIFFERENT DATES FROM HISTORIANS-609, 606, 588, 586, AND SO ON AND ON? The answer is that HISTORIANS HAVE HANDED US DATES THAT HAVE MISLED US FOR YEARS, dates based on deficient human science of dating, by which they have dated the earth and the universe in the millions and billions of years. This is one of the reasons Ussher's chronological studies are faulty. He had employed one of these dates from the historians, the 586 B.C. date. Every other theologian that has undertaken the Biblical Chronology study has made the same mistake, including William Miller.

THE DISCREPANCY OF THE HISTORIANS' DATE WITH PROPHECY AND WHY IT MISLED BOTH USSHER AND WILLIAM MILLER

Let us take the earliest of the dates of the historians, 586 B.C., which is the date Ussher used for his chronology, and William Miller used to interpret Daniel 8:14 prophetic period of the 2300 days. If we take this date to be the date Jerusalem fell to Nebuchadnezzar, thus starting off the 70 years of captivity, it will mean that the seventy years will end in **516 B.C.** Since according to Jeremiah, the decree to end captivity must come exactly 70 years to the day, it will mean that Cyrus' decree was given in 516 B.C. But this must have presented a problem to William Miller, because the **SEVENTY WEEKS (490 years)** determined on the post-exilic nation of Israel(Daniel 9), will not even reach down to the date Jesus was born, and therefore will be out of sync with the Daniel 9 events.

Yet we know that Daniel 9 and Daniel 8 are linked, and therefore the decree to build and restore Jerusalem must line up with the last prophetic week of Jesus' ministry. Since the dates of the last seven years of Jesus' ministry (27, 31, and 34 C.E.) are authenticated, it means that for this to be the case, the 490 years must begin in 457 B.C. So this date is unquestionably accurate and reliable. We know from prophecy that Isaiah, Jeremiah and Daniel together indicated that Daniel 9:25 marks the end of the 70 years of captivity, which means it is Cyrus's decree that fulfills Daniel 9:25. But based on this 586 B.C. date given by historians, the 457 B.C., is a whopping **59 years** after the 70-year captivity had ended. Thus, there is a discrepancy between historians' date and the prophecies of Isaiah, Jeremiah and Daniel, regarding when the decree to build and restore Jerusalem actually was declared. Who is correct, the historians or the prophets? To anyone who believes God and scriptures, the answer should be obvious, and that is that the prophecies are right, and the historians are wrong.

THE ARTAXERXES DECREE ARGUMENT WHICH WILLIAM MILLER FALSELY LATCHED UNTO

Faced with the above-mentioned discrepancy, and in an attempt to harmonize scriptures with the dates of the historians, William Miller, and the advent movement he championed, came up with an argument that contradicts Isaiah, Jeremiah and Daniel's prophecies, namely, by stating that it is Artaxerxes's decree that fulfilled Daniel 9:25, based on their reliance and acceptance that the historian's dates are correct. Since 457 B.C., is the date that syncs with Daniel 9 events and dates, and believing the historians' dating of Artaxerxes decree to be given in 457 B.C., William Miller and his followers resolved therefore that the only way to resolve the dilemma is to see Artaxerxes decree as the correct decree that fulfilled Daniel 9:25, instead of Cyrus decree. Their argument is encapsulated in the following excerpts from Uriah Smith's ***Daniel and Revelation***.

"The Decrees of Cyrus, of Darius and of Artaxerxes

There are but four events which can be taken as answering to the commandment to restore and build Jerusalem. These are, (1) The decree of Cyrus for the rebuilding of the house of God, B.C.536 (Ezra1:1-4); (2) The decree of Darius for the prosecution of that work, which had been hindered, B.C.519 (Ezra6:1-12); (3) The decree of Artaxerxes to Ezra, B.C.457 (Ezra7); and (4) The commission to Nehemiah from the same king in his twentieth year, B.C.444. Nehemiah 2. Dating from the first two of these decrees, the seventy weeks, being weeks of years,[1] 490 years in all, would fall many years short of

reaching even to the Christian era; besides, these decrees had reference principally to the restoration of the temple and the temple-worship of the Jews, and not to the restoration of their civil state and polity, all of which must be included in the expression, "To restore and to build Jerusalem." <u>These made a commencement of the work. They were preliminary to what was afterward accomplished. But of themselves they were altogether insufficient, both in their dates and in their nature, to meet the requirements of the prophecy; [223] and thus failing in every respect, they cannot be brought into the controversy as marking the point from which the seventy weeks are to date</u>. The only question now lies between the decrees which were granted to Ezra and to Nehemiah respectively".

The very first flaw with this argument is that it is based on the dates provided by historians which do not, almost always, agree with bible's view on when these events actually occurred.

The second flaw is that Uriah Smith and William Miller of course, presumed that Cyrus' decree does not qualify as the decree that fulfills Daniel 9:25 because it does not explicitly include the wordings that convey **"restoration of civil state and polity"** of the nation., but only talks about restoring the temple. But does Daniel 9:25 require this specification that Uriah Smith advances? Doesn't the rebuilding of the temple obviously indicate and warrant that the city is also to be rebuilt, and the polity restored? How can the temple be built without the city built? Does this even make any sense? On what ground is the temple to be built? Is it not on the grounds of Jerusalem? Is Cyrus so ignorant of the relationship of the temple to the city and nation that he should mean his decree for only the temple and not the city? To assume this amount to ignorance or a desperate attempt to support a wrong preconceived idea. Isaiah's prophecy clearly indicate that Cyrus' decree is a complete and comprehensive decree that authorizes the rebuilding of city and temple. Here is Isaiah's prophecy:

"That(is God) saith of Cyrus, He is my shepherd, and shall perform all my pleasure, even saying to Jerusalem, Thou shalt be built; and to the temple, Thy foundation shall be laid"

There we have it. The preposterous ide and defense offered for why Cyrus decree was not sufficient to fulfill Daniel 9:25, is here dismantled completely by Isaiah's prophecy. Even without this plain statements from Isaiah's prophecy, common sense alone will tell one that it is not sensible to ask the people to just build the temple and not build up the city. Cyrus simply does not see any reason for his decree that authorizes the temple to be rebuilt as not obviously indicative that the city can also be rebuilt. A learned person does not have to waste words on what is obvious. So, Cyrus knew that his decree automatically should be see as including that the city be rebuilt also. Otherwise of what purpose and use is the decree to the people? Will the temple float in the air? Will the people float in the air while worshipping and ministering in the temple? Therefore the argument for why Cyrus decree is not the one that fulfills Daniel 9:25, is deficient at best.

Let us read Daniel 9:25 again:

"Know therefore and understand, that from the going forth of the commandment to restore and to build Jerusalem unto the Messiah the Prince shall be seven weeks, and threescore and two weeks: the street shall be built again, and the wall, even in troublous times".

Note the phrase, **"from the going forth".** It can be seen that nothing in this expression precludes further additional decrees that expand and reaffirm the original decree from being given. But let us read Cyrus' decree and assess what it mandates. It reads:

Ezra 1:1-11

"Now in the first year of Cyrus king of Persia, that the word of the LORD by the mouth of Jeremiah might be fulfilled, the LORD stirred up the spirit of Cyrus king of Persia, that he made a proclamation throughout all his kingdom, and put it also in writing, saying, [2] Thus saith Cyrus king of Persia, **The Lord God of heaven hath given me all the kingdoms of the earth; and he hath charged me to build him an house at Jerusalem, which is in Judah. [3] Who is there among you of all his people? his God be with him, and let him go up to Jerusalem, which is in Judah, and build the house of the Lord God of Israel, (he is the God,) which is in Jerusalem. [4] And whosoever remaineth in any place where he sojourneth, let the men of his place help him with silver, and with gold, and with goods, and with beasts, beside the freewill offering for the house of God that is in Jerusalem. [5]** Then rose up the chief of the fathers of Judah and Benjamin, and the priests, and the Levites, with all them whose spirit God had raised, to go up to build the house of the Lord which is in Jerusalem. [6] And all they that were about them strengthened their hands with vessels of silver, with gold, with goods, and with beasts, and with precious things, beside all that was willingly offered. [7] Also Cyrus the king brought forth the vessels of the house of the Lord, which Nebuchadnezzar had brought forth out of Jerusalem, and had put them in the house of his gods; [8] Even those did Cyrus king of Persia bring forth by the hand of Mithredath the treasurer, and numbered them unto Sheshbazzar, the prince of Judah. [9] And this is the number of them: thirty chargers of gold, a thousand chargers of silver, nine and twenty knives, [10] Thirty basons of gold, silver basons of a second sort four hundred and ten, and other vessels a thousand. [11] All the vessels of gold and of silver were five thousand and four hundred. All these did Sheshbazzar bring up with them of the captivity that were brought up from Babylon unto Jerusalem".

Now, it is true that Cyrus decree does not have the specific words and phrases about ***civil and political administration***, but the important question is: Does the decree forbid the people from establishing the civil and political structures of the state? Did the people need to be instructed explicitly, especially by a non-Jew, on how to administer their state, whether spiritually, civilly, or politically? It is noteworthy that Cyrus did not instruct in the decree on how they were to perform the religious rites, yet as soon as they arrived, the people established and began the religious rites, because they knew that they did not even need anything to be built before they could do that.

So, the fact that they did not have the civil laws and administration fully in place is not because Cyrus decree did not spell it out in the decree, but because they had to do first things first. Cyrus decree therefore never forbade them from setting up any aspect of the nation's administration, and as such cannot be said to be incomplete in the fulfillment of Daniel 9:25. The people did not need to be told how and when to institute the instruments of administration. All they needed was just the decree that they could go back and re-establish the self-governing state.

Once again, because this cannot be overstated, let us ask the question: Does it make any sense that a decree to rebuild the temple also excludes the administration of the temple rites and governing of the people or exclude the rebuilding of the entire city? The life of the nation is actually the temple and its services. It has to be remembered that God is still the Chief ruler of this nation, at least in the official sense. Cyrus knew this, and so did not need to spell out civil and political terms for them. Those terms were for the people to determine and establish, which they had not forgotten because they are in the book of the covenant.

However, let us see, based on Uriah Smith's argument, if Artaxerxes decree actually does what Cyrus decree did not set in motion. Artaxerxes decree reads as follows:

Ezra 7:11-27

"Now this is the copy of the letter that the king Artaxerxes gave unto Ezra the priest, the scribe, even a scribe of the words of the commandments of the Lord, and of his statutes to Israel. [12] Artaxerxes, king of kings, unto Ezra the priest, a scribe of the law of the God of heaven, perfect peace, and at such a time. [13] I make a decree, that all they of the people of Israel, and of his priests and Levites, in my realm, which are minded of their own freewill to go up to Jerusalem, go with thee. [14] Forasmuch as thou art sent of the king, and of his seven counsellors, to enquire concerning Judah and Jerusalem, according to the law of thy God which is in thine hand; [15] And to carry the silver and gold, which the king and his counsellors have freely offered unto the God of Israel, whose habitation is in Jerusalem, [16] And all the silver and gold that thou canst find in all the province of Babylon, with the freewill offering of the people, and of the priests, offering willingly for the house of their God which is in Jerusalem: [17] That thou mayest buy speedily with this money bullocks, rams, lambs, with their meat offerings and their drink offerings, and offer them upon the altar of the house of your God which is in Jerusalem. [18] And whatsoever shall seem good to thee, and to thy brethren, to do with the rest of the silver and the gold, that do after the will of your God. [19] The vessels also that are given thee for the service of the house of thy God, those deliver thou before the God of Jerusalem. [20] And whatsoever more shall be needful for the house of thy God, which thou shalt have occasion to bestow, bestow it out of the king's treasure house.

[21] And I, even I Artaxerxes the king, do make a decree to all the treasurers which are beyond the river, that whatsoever Ezra the priest, the scribe of the law of the God of heaven, shall require of you, it be done speedily, [22] Unto an hundred talents of silver, and to an hundred measures of wheat, and to an hundred baths of wine, and to an hundred baths of oil, and salt without prescribing how much. [23] Whatsoever is commanded by the God of heaven, let it be diligently done for the house of the God of heaven: for why should there be wrath against the realm of the king and his sons? [24] Also we certify you, that touching any of the priests and Levites, singers, porters, Nethinims, or ministers of this house of God, it shall not be lawful to impose toll, tribute, or custom, upon them. [25] And thou, Ezra, after the wisdom of thy God, that is in thine hand, set magistrates and judges, which may judge all the people that are beyond the river, all such as know the laws of thy God; and teach ye them that know them not. [26] And whosoever will not do the law of thy God, and the law of the king, let judgment be executed speedily upon him, whether it be unto death, or to banishment, or to confiscation of goods, or to imprisonment".

It could be seen that the Artaxerxes decree says nothing about building and restoring, but was mostly about the administration of the Law. ***THE REBUILDING OF JERUSALEM WAS ALREADY UNDERWAY before Artaxerxes decree. That decree does not even say anything about building. It only relates to the ADMINISTRATION OF THE LAW.*** As pointed out earlier, there is nothing in Cyrus decree that would have precluded them from administering the Law. The task of rebuilding the temple and maintaining order among the people who were already settled in their various tribes, and lots would have required administration and leadership. They already had leaders who were administering law and order before the Artaxerxes decree. The idea that Cyrus decree was limited in its scope is therefore false.

So, what essentially was Artaxerxes decree about? It was simply **A REINFORCEMENT OF CYRUS DECREE,** just as that of Darius before him. To say that Daniel 9:25 never became fulfilled until Artaxerxes decree is false and misleading. How can a decree that asks a people to go and rebuild their Temple, the very object that is central to their existence, and government, be seen as precluding their self-government and autonomy? Such an idea is absurd. The decree asking for the temple to be rebuilt automatically embraces and authorizes the rebuilding of the city, since, after all, the temple is a central structure of the city.

Therefore, Cyrus' decree is ***THE PRIMARY DECREE***. It can be likened to God's everlasting covenant. As Paul noted, it was CONFIRMED BEFORE OF GOD IN CHRIST, IN HEAVEN, on the very day Adam sinned. But through the chosen SEEDS (Adam, Noah, Abraham, to mention a few), God re-affirmed the covenant over the years until the CHIEF SEED himself, Jesus Christ came to earth and ratified it with his death. Likewise, Cyrus decree was the primary decree while the others were simply **ADJUNCTS** that **REAFFIRMED, REAUTHORIZED AND REINFORCED** the original decre, which was necessitated because the Jews were not making good progress, mostly because of opposition ("TROUBLOUS TIMES") that Daniel prophetically saw-**Daniel 9:25 KJVS:**

"the street shall be built again, and the wall, even in troublous times".

IN CONCLUSION: Cyrus decree did fulfill Daniel 9:25, and is the primary decree. It is therefore the decree that should have the date that fulfills the Daniel 8 and 9 prophetic timeline. Consequently, it is the decree that should have the date of 457 B.C., in spite of any historical dates that historians have provided, which are at best, guesswork based on false research and false science of carbon dating and the imperfect science of archaeology. When Cyrus decree is assigned its rightful date of 457 B.C., we therefore have the beginning of the 70 years of captivity as 527 B.C. With this date of 527 B.C., we have the right date for the chronology of the earth and Great Controversy, the correct Gregorian date on which the biblical part of the chronology is to be anchored.

Isaiah, Daniel, Ezra, Nehemiah, Jeremiah, all make it clear that the decree to rebuild and restore Jerusalem unto the Messiah was to be given exactly at the 70TH year, and that that decree was CYRUS' DECREE. They all made it clear that it is Cyrus' decree, not Artaxerxes' decree, that fulfills Daniel 9:25, and that this decree should lead up to and line up with when Jesus was baptized, crucified, and Stephen killed. But based on the date of 586 B.C., by the historians, William Miller found out that he could not reconcile this apparent discrepancy. So, still relying on historians instead of scripture, the same mistake he made in relying on the theologians of his day on the meaning of sanctuary instead of scriptures, he decided to go along with the error that 457 B.C. belongs to Artaxerxes decree, and erroneously chose Artaxerxes decree instead of Cyrus decree as the one that fulfilled Daniel 9:25.

Thus relying on historians and their dating that are based on archeology, misled William Miller, not with regards to the 457 B.C. date as that date is correct, but in wrongly assigning the date to the wrong decree and wrong king, Artaxerxes, thereby involving the chronology of the earth in error. Since the it is Cyrus' decree that fulfills Daniel 9:25, and 457 B.C. is the correct date that explains Daniel 9 70 WEEKS, it therefore means that the 457 B.C. date belongs to Cyrus' decree and not to Artaxerxes decree. As such the 70 years captivity began in 527 B.C.

Now that we have exhausted the flaws in Ussher's study, let us now present the accurate chronology of the earth and the Great Controversy.

CHRONOLOGY AND AGE OF THE EARTH

FROM ADAM TO EXODUS IS **2728 YEARS (**Different from Ussher's due to what he chose as starting date of the 430 years sojourn in Egypt**)**

FROM EXODUS TO WHEN THE BUILDING OF THE FIRST TEMPLE BEGAN (4TH YEAR OF SOLOMON'S 40-YEAR REIGN) IS 479 YEARS (1Kings 6:1)

FROM THE ABOVE TO THE END OF SOLOMON'S REIGN IS 37 YEARS (1Kings 11:42).

FROM END OF SOLOMON'S REIGN TO NEBUCHADNEZZAR DESTROYING JERUSALEM IS 390 YEARS (See Ezekiel 4: 4-6).

THEN ALL THE ABOVE ARE JOINED TO 527 C.E WHICH IS WHEN NEBUCHADNEZZAR DESTROYED JERUSALEM (The reason this date is chosen as against Ussher's 588 C.E., has been discussed earlier).

SO, WE HAVE: 2728 + 479 + 37 + 390 + 527 = 4161 C.E.

SO, 4161 C.E IS THE YEAR THE EARTH AND ADAM WERE CREATED (ADAM'S CREATION BEING ON THE SIXTH DAY).

THAT MAKES THE YEAR THE EARTH REACHED 6000 YEARS TO BE AD 1839.

SO, BY OUR PRESENT YEAR AD 2025 THE WORLD IS 6186 YEARS.

IN AD 2027 THE WORLD WILL BE 6188 YEARS

IN AD 2031 THE WORLD WILL BE 6192 YEARS.

IN AD 2034., THE WORLD WILL BE 6195 YEARS.

IN AD 2038 THE WORLD WILL BE 6199 YEARS.

IN AD 2042, THE WORLD WILL BE 6203 YEARS.

Based on this accurate chronology,

AT **JESUS' BIRTH IN 2 BC**, THE EARTH WAS **4159 YEARS** OLD.

AT JESUS' BAPTISM IN **AD 27**., THE EARTH WAS **4188 YEARS** OLD.

AT JESUS CRUCIFIXION IN **AD 31.,** THE EARTH WAS **4192 YEARS** OLD.

IN **AD 34** WHEN THE 490 YEARS ALLOTTED TO THE JEWISH NATION ENDED, THE EARTH WAS **4195 YEARS** OLD.

IN AD 66 WHEN THE ABOMINATION OF DESOLATION (ROMAN POWER) ENCOMPASSED JERUSALEM, THE EARTH WAS **4227 YEARS** OLD.

IN AD 70, WHEN THE DESTRUCTION OF JERUSALEM OCCURRED, THE EARTH WAS **4231 YEARS** OLD.

All of these events therefore occurred within the range of approximation to 4000 YEARS, which verifies the Spirit of prophecy (by Ellen G. White's) 4000 years prophetic statements concerning the ***birth, baptism***, and ***death*** of Jesus. Later in this study, in the second section, more will be said in regard to the relation of chronology of the earth to Ellen G. White's statements on the age of the Great Controversy. But if we assume the age of the earth to be approximately equal to the age of the Great Controversy, in other words Adam sinned in the first year of creation, we can see that Ellen White's prophetic statements that the conception(and birth), baptism, and death of Jesus all occurred ***about*** 4000 years is authenticated by scriptures.

AGE OF THE GREAT CONTROVERSY-WERE ELLEN G. WHITES 4000, 6000 YEARS EVER MEANT TO BE EXACT OR JUST APPROXIMATE?

Her prophetic statements related the 4000 years figure to the BIRTH, BAPTISM, and DEATH OF CHRIST, which have the dates 2 C.E, AD 27, and AD 31, respectively. The prophecy is also based on when the Great Controversy began. Based on Ussher's chronology, many Adventists have discovered that it is impossible to align it with her prophecy. But as we have shown, Ussher's chronology is flawed. That notwithstanding, even with the chronology done in this study, it is also impossible to get the 4000 years to align with the birth, baptism or death of Christ. We will demonstrate that "the math is simply not there", as is often said.

To get the age of the great controversy (which starts from when Adam sinned), we employ the equation for the age of the earth, by subtracting Adam's age when he sinned from the section of the chronology- creation to exodus. But we have to know what Adam's age was when he sinned.

There is no scripture that expressly give us Adam's age when he sinned, or Abel's age when Cain killed him, both information that we need to compute the age of the Great Controversy.

To proceed therefore, we will have to make Adam's age when he sinned the **UNKNOWN FACTOR- (X)**.

Thus for the birth of Jesus, which occurred in 3 BC, to be at the 4000th year of the GC, we will have an equation that looks as follows:

(X) = 4000 - (2728 + 479 + 37 + 390 + 527 – 2)

Solving for **(X),** we will have:

(X) = 4159 - 4000

(X) = 159

Thus, Adam would have to be 159 years old when he sinned for the GC to be 4000 years at Christ's birth. This as can be seen is not plausible as it would mean that Dam gave birth to Seth well before he had sinned. That would make Seth's conception and birth to be immaculate or sinless. But scriptures are clear that Adam had Seth after Abel's death at which time Adam and eve had already fallen from grace and sinned and been driven out of Eden.

For the baptism (AD 27) to have occurred at the 4000th year of the GC, the equation will be as follows:

(X) = 4000 – (2728 + 479 + 37 + 390 + 527 + 27)

(X) = 4188 - 4000

(X) = - **<u>188 years</u>** (This will be Adam's age when he sinned in order for the 4000 years to terminate in 27 A.D.)

As can be shown from this equation, it is impossible for Adam to have sinned at this age. This is because Abel would only be 2 years old when he was killed by Cain (Recall that Seth was born after Abel had been killed).

For the date of Jesus' crucifixion, AD 31, to be the 4000th year of the GC, the equation will be as follows:

(X) = 4000 – (2728 + 479 + 37 + 390 + 527 + 31)

(X) = 4192 – 4000 = **<u>192 years</u>** (Adam's age when he sinned in order to get the year 31 AD to be the 4000th year of the GC.

As can be seen, all these ages do not even make room for Abel to have existed. Moreover it means that Seth was born well before Adam had sinned, which of course is not the case.

THE IMPORTANT CONCLUSIONS: Therefore, no mathematical scenario makes it possible for the birth, baptism, or death of Christ to have corresponded with the 4000th year of the GC. Therefore Ellen G. White's prophecy about the 4000, 6000 years were given to be approximate and not exact, reminiscent of God telling Abraham that her seed will be afflicted four hundred years, even though the exact period, as determined in this study so far, was **359 YEARS.**

Also in conclusion from all these, the second coming of Jesus **WILL NEVER** be at the 6000TH year of earth's creation, **NEITHER WILL** it be at the 6000TH year of the Great Controversy because those dates are all past.

Thus, the theology of the **SEVEN-THOUSAND-YEAR REDEMPTION CALENDAR,** based on the creation week analogy and extrapolation, is not supported by scriptures. Peter and David's statements about a thousand year being like a day in the eyes of God, and vice versa, are simply alluding to the fact that God's perception of time is different from ours, and that God is above and beyond time.

This also means that Ellen G. White's 6000 years statements on the second coming of Jesus were an approximation. Those statements are saying to us that on the day Jesus will come, the world would have been only **ABOUT 6000 YEARS**. This study reveals just how close to the sixth thousandth year Jesus' coming will be, in validation of the inspiration of Ellen G. White's messages.

THE TRUE PURPOSE OF ELLEN G. WHITE'S 4000, 6000 YEARS PROPHETIC STATEMENTS

THE MISAPPLICATION OF ELLEN G. WHITE'S (SPIRIT OF PROPHECY) 4000, 6000 YEARS MESSAGES: The popular application that has been made of Ellen G. White's 4000, 6000 years statements, is that Jesus's second coming is expected to occur at the 6000th year of the GC. Those who make this application are also mostly believers and teachers of the doctrine of the SEVEN THOUSAND YEARS REDEMPTION CALENDAR, that is patterned after the CREATION WEEK. This pattern presumes that the seven days of the creation week represent seven thousand years in which God will conclude the business of sin and salvation: six thousand years of sin, and one thousand years of rest in heaven. The attempt to support this from scriptures is extremely weak and contrary to rational principles of biblical knowledge. It is based on arbitrary theories and extrapolations.

Let us examine the premises on which this doctrine is built. The most important and foundational is the ONE YEAR EQUALS ONE THOUSAND YEARS theory. This theory and principle is falsely drawn from the Psalmist's and Peter's statements about a thousand year being as one day in the eyes of God and vice versa. Let us read those scriptures.

"Before the mountains were brought forth, or ever thou hast formed the earth and the world, even from everlasting to everlasting, thou art God. Thou turnest man to destruction; and sayest, Return ye children of men. For a thousand years in thy sight are but as yesterday when it is past, and as a watch in the night" (Psalms 90:1-4)

"But beloved, be not ignorant of this one thing, that one day is with the Lord as a thousand years, and a thousand years as one day" (2 Peter 3:8)

Let us begin with the Psalmist statement. One can easily and quickly see here that the text is simply explaining the term, "everlasting". To humans, a thousand years is a very long stretch of time, and in fact no human ever lived to be a thousand years old. The psalmist is saying that this is like yesterday, meaning, like nothing. The idea is that God is above time because what we humans consider as time, both in the sense of past, present, and future, is simply an eternal continuum with God. Time was created only for humanity and this world. The psalmist is simply extolling and declaring and explaining the greatness, limitlessness, and timelessness of God. There is no suggestion in the above texts that God is establishing a principle of time for Himself that means ONE THOUSAND YEARS IS EQUAL TO ONE DAY. If that were the case, why choose only one thousand years? Why not one million, one billion, or even one trillion? Wouldn't it still be equally true?

As for Peter's reference to the Psalmist's words, notice that Peter did a vice versa with the statement. In other words Peter is not setting a principle or rule. Peter is simply employing a figure of speech, just like the Psalmist, to indicate that God is beyond and above time. The fact that Peter did a vice versa with the statement is a clear indication that he was not declaring a rule.

ONE DAY IS LIKE ONE THOUSAND YEARS TO GOD

ONE THOUSAND YEARS IS LIKE ONE DAY TO GOD.

The two statements cancel each other. Mathematically, it means two equations that cancel, which equals to zero, again corroborating the Psalmist's view that there is zero time with God. No rule or law of time applies to God. He lives outside of time. But theologians decided to make this about humans, and chose just one direction of the above statements: ONE DAY EQUAL TO ONE THOUSAND

YEARS, and ignore the reverse statement. Those who have employed Peter's text to support the creation week being equivalent to seven thousand years, do so under the presumption and false premise that the CREATION WEEK applies to God, that God is regulated by time as humans are. But the creation week is a time indices created for humans, not for God. God is not bound by it, because He lives above it.

How can God rest for just one 24-hour day? What does He do with the rest of His eternity? This simply goes to show that this 24-hour rest day is not for Him but for humanity. Time is created by God for humanity and not for Himself. Peter is simply saying that while humanity bases his actions and expectation on time, God does so solely on the events He seeks to accomplish, which in this case is completing the number of people that will populate his kingdom and letting Satan and humans complete their own acts in the conflict.

Based on Peter's text, those who employ it to say CREATION WEEK IS EQUAL TO SEVEN THOUSAND YEARS, are also obligated and bound to say SEVEN THOUSAND YEARS IS EQUAL TO CREATION WEEK, thus cancelling out their erroneous conclusion.

SEVEN DAYS (CREATION WEEK) = SEVEN THOUSAND YEARS (To God, not to man)

SEVEN THOUSAND YEARS = SEVEN DAYS (CREATION WEEK) (To God, not man).

Both are true at the same time. So, it all cancels out, and the net fact is zero time. No time. Therefore, the SIX THOUSAND YEARS OF SIN-ONE THOUSAND YEARS OF REST-THEOLOGY is not based on scriptures. This explains why Ellen G. White never believed and never taught it. One important evidence that Ellen G. White never associated her 4000, 6000 years messages with the SEVEN THOUSAND YEARS THEOLOGY, is that **SHE**

HONESTLY AND EARNESTLY EXPECTED JESUS TO COME IN HER DAY.

Alternatively, she employed the 4000 years ***APPROXIMATELY*** in relation to the three events(Jesus's birth, baptism, and death), and knowing that by rule of approximation, anything from 3500 years to 4499 years can still be approximated to 4000 years, she could therefore rightly expect Jesus to come in her day, and it would still have been about 4000 years. This explains why she employed such qualifiers as ***ABOUT, NEARLY, OVER***, in her statements.

In developing the seven thousand years redemption calendar, the proponents claim that the millennium will allow the earth to rest and keep the sabbath. But what they forget is that after the millennium, nothing of this earth will come to remembrance. It will all be dissolved and clean removed, and a new heaven and earth created(Revelation 21). The plan is never for this earth and heaven to rejuvenate, the idea that the leaders of the earth are advocating presently. God has a completely different plan, which is to completely erase what is now and replace it with something entirely new, just as He will give the saints a completely new body that will inhabit it. Therefore the idea of sabbath millennial rest for the earth is redundant and meaningless.

DOCTRINE AND KNOWLEDGE OF THE TWO LAST GENERATIONS-THE JEWISH AND THE WORLD, AND HOW THEY ARE LINKED BY ELLEN G, WHITE'S 4000, 6000 YEARS PROPHETIC TIMES.

Both Christ and Daniel directly speak to the two last generations, the last generation of the Jewish nation that ended AD 70 and that of the world that will end with the second coming of Jesus, thus making the destruction of Jerusalem a TYPE of the end of the world when Jesus comes. This is a very crucial point, especially in understanding Ellen G. White's 4000, 6000-year-old prophecy.

Speaking to the two last generations, Jesus declared:

"Verily, I say unto you, this generation shall not pass, till these things be fulfilled" (Matthew 24:34)

While giving this prophecy on the last generations, Christ however clearly made the important connection with the prophecies of Daniel:

"When ye therefore shall see the abomination of desolation, spoken of by Daniel the prophet, stand in the holy place, (whoso readeth, let him understand:)".

The fact that Jesus did not sectionalize his prophecy is a clear indication that he meant it to have dual application, and the fact that he referenced back to the prophecies of Daniel meant that the books of Daniel reveal the truth about the two last generations. What this means therefore is that Jesus is citing the book of Daniel as a resource material from which we can determine what part of his prophecy is about the last generation of the Jewish nation and which part is specific to the last generation of the earth, and which part applies to both. Therefore, let us go to the book of Daniel for more specifics about the two last generations.

LAST GENERATION OF THE JEWISH NATION

Daniel 9:24-26 KJVS

"Seventy weeks are DETERMINED UPON THY PEOPLE AND UPON THY HOLY CITY, to FINISH THE TRANSGRESSION, and to make an end of sins, and to make reconciliation for iniquity, and to bring in everlasting righteousness, and to seal up the vision and prophecy, and to anoint the most Holy".

[25] KNOW THEREFORE and understand, that from the going forth of the commandment to restore and to build Jerusalem UNTO THE MESSIAH THE PRINCE shall be seven weeks, and threescore and two weeks: the street shall be built again, and the wall, even in troublous times".

We could easily read these texts and yet not grasp the important message revealed in them. Here God is saying to the nation through Daniel, "You are getting one last period for your existence, and that will be 70 weeks, because I need to send my Son to fulfill, confirm, and seal the new covenant with His death, with his blood". This prophecy, though conditional on the nation's obedience or disobedience to the covenant, is also certain to fulfill based on God's sovereignty(foreknowledge, wisdom and strength). The balance between humanity's choice and His sovereignty is something only God can fathom. If Israel had accepted God's son, she would never have been destroyed by Rome. In fact the possibility was there that their hope to be rid of Roman rule would have been realized. Christ could have honored her with at least this much, if not entirely reversing the fate of Rome by bringing her under the rule of Israel.

So, God is saying in plain terms that the nation is coming back for another existence of 70 weeks of years (490 years) to complete its destiny, which is primarily to give His Son the place and circumstance to confirm the New Covenant, seal and establish the Plan of Salvation. The statement, ***"to finish the transgression",*** is very significant. Notice the use of definite article, THE, a clear indication that there is a particular transgression being referred to. Earlier in Daniel 8:, it is called TRANSGRESSION OF DESOLATION. This transgression is the kind that renders the transgressor SPIRITUALLY DESOLATE, in other words, beyond redemption. It closes probation for the entity in question. Applied to a people or nation, it is the transgression, or the extent of it, that closes their probation. Put in another term, the nation has FILLED THE CUP OF HER INIQUITY. This reference was used for the land of Canaan before God brought Israel from Egypt to overrun and occupy it. Another way this transgression is also described is as ***TAKING AWAY THE DAILY (GOD AND HIS DOCTRINE OF GRACE AND OBEDIENCE) and REPLACING THIS WITH THE ABOMINATION OF DESOLATION DOCTRINE.***

Essentially therefore, God revealed to the nation, through Daniel, that the COMING OF HIS SON WILL MARK THE BEGINNING OF THE LAST GENERATION OF THE NATION. It is on the back of this knowledge that Jesus gave the disciples the important message,

Matthew 24:34 KJVS

"Verily I say unto you, THIS GENERATION shall not pass, till all these things be fulfilled".

Thus Jesus was saying clearly, “Look the nation was brought back from Babylon, not because of the righteousness of the people, but for Me, so I can accomplish the plan of salvation: to establish and seal it with my blood". In other words, Jesus is the reason the nation was brought back from captivity. The nation clearly did not appreciate this fact and truth, for otherwise they would have embraced and held on to Jesus dearly. The one person that would have ensured that their nation remained and not destroyed, they ignorantly and wickedly TOOK AWAY AND REPLACED WITH CAESAR. To Pilate's query, "Shall I crucify your king", a question that is also the same as, "Shall I crucify your God, the Head of your religion, your High priest?", they answered, "We have no king but Caesar", "If thou let this man go, thou art not Caesar's friend"(John 19:12, 15). In their spiritual dearth, ignorance, and obsession with killing their TRUE KING AND MESSIAH, they TOOK AWAY THE DAILY AND SET UP ABOMINATION OF DESOLATION(Caesar, Rome, and the Sun god of Rome in their sanctuary). Indeed, as Christ declared to them, their HOUSE(SANCTUARY) was left unto them **SPIRITUALLY DESOLATE**. And that which is spiritually desolate is ripe and ready for eventual **PHYSICAL DESOLATION**.

It stands to reason also that if Jesus was the reason the nation was brought back, their rejection of Jesus will therefore surely guarantee that the nation will be lost again. This, as history testifies, in agreement with prophecy, was what happened; the nation rejected Jesus, and Rome eventually destroyed the nation. It brings into question the matter of the coming back of the nation in 1948 A D. With the nation still in her rejection of Jesus as Messiah and King, could God be behind their return in 1948 A.D.? The answer is, absolutely no. This therefore means that Satan is behind this return of Modern nation of Israel. This being the case, Jesus' prophecy in Luke still stands, and will yet be fulfilled again:

"And they shall fall by the edge of the sword, and shall be led away captive into all nations: and Jerusalem shall be trodden down of the Gentiles, UNTIL THE TIMES OF THE GENTILES be fulfilled" - Luke 21:24 KJVS

BACK TO THE SUBJECT OF THE LAST GENERATION: Therefore the coming (conception and birth) of Jesus clearly marked the BEGINNING OF THE LAST DAYS, LAST GENERATION, OF THE NATION. Recall Paul calling the days of Christ's first advent THE END OF THE WORLD(Hebrews). In Matthew 1, after Jesus had gone back to heaven, Matthew revealed this fact even more plainly in his genealogy chronicle, From Abraham to Jesus. Why from Abraham? Because the nation of Israel began and came from Abraham. In AD 34, the "70 WEEKS" (490 years) allotted to the Jewish nation came to a close, and with it the close of probation for the nation. The message for salvation from then on was for individual Jews to save themselves from the doomed nation. The message, "Come out of her my people, did ring true for the people of the nation and church. When the church received and accepted Caesar in the place of her rightful God, King and High priest, they TOOK AWAY THE DAILY AND SET UP ABOMINATION OF DESOLATION, and thus became a FALLEN CHURCH. This was exactly to happen to the Apostolic church also later, when ultimately she, in union with Roman state, replaced Christ and God in the church with another Caesar(in sacerdotal cloth)-the Pope of Rome, as Head, King, God of the church leading to the emergence of that evil power called PAPACY.

THE LAST GENERATION OF THE WORLD

Because Christ's prophetic discourse IN Matthew 24 had dual application, Jesus was therefore equally prophesying about the LAST GENERATION OF THE GREAT CONTROVERSY AND END OF THE WORLD. It means that the important Great Controversy events of the last generation of the Jewish nation ***will be repeated*** in the last generation of the world. What are these events? Daniel 8 and 9 identify these as follows:

1.BEGINNING OF THE GATHERING OF THE LOST SHEEP (JESUS BEGAN THIS AS SOON AS HIS BAPTISM IN A.D. 27).

2.*TAKING AWAY OF THE DAILY* (REJECTION OF JESUS AND HIS DOCTRINE AND PRIESTHOOD AND CRUCIFYING HIM) AND ACCEPTING THE ABOMINATION OF DESOLATION (CAESAR) AS KING OF THE CHURCH AND STATE OF ISRAEL.

3.THE CLOSE OF PROBATION.

4. *ABOMINATION OF DESOLATION* STANDING IN THE HOLY PLACE; AND FINALLY THE DESTRUCTION.

In Daniel 12, Daniel was shown that the above events will repeat in the last generation of the world leading to the STANDING OF UP OF JESUS (AS MICHAEL) and his SECOND COMING. BUT how do we know when the last generation of the Great Controversy begins? WHAT IS THE TIME-LINK BETWEEN THE TWO LAST GENERATIONS? This is where ELLEN G White's 4000, 6000 years prophecies come in. This is the actual purpose of the messages. They provide the 2000-year-old link between the two last generations, especially between the important and major events of that generation and the parallel of those events in the last generation of the world. The purpose was never to pin down the second coming at the 6000th year of the Great Controversy. Through Spirit of Prophecy messages, God revealed to E.G. White that the conception and birth of Jesus occurred about the 4000TH YEAR OF THE GREAT CONTROVERSY.

In A.D. 34, it is understood that PROBATION ENDED FOR THE NATION OF ISRAEL. God therefore was telling His beloved servant that 2000 years later from when the 70 weeks of years were completed for the Jewish nation, will be the END OF PROBATION FOR THE WORLD. This will therefore be in the year 2034. The 6000th year of the Great Controversy was therefore never given to her as the definite year Jesus was to come, but rather the year about which the end will come, meaning that by the time the end will come, it would only have been about 6000 years since Adam sinned.

THE EVENTS OF THE TWO LAST GENERATIONS (ISRAEL AND THE WORLD) COMPARATIVELY PAIRED BY THE PROPHETS-DANIEL, JOHN, PAUL, AND ELLEN G. WHITE.

From Ellen G. White we know the date for the end of probation for the world, and that is OCTOBER 22 A.D. 2034. This is when the INVESTIGATIVE JUDGEMENT that began in OCTOBER 22, 1844, will end, and the EXECUTIVE JUDGEMENT begins with the SITTING OF ANCIENT OF DAYS. ***DOES THIS CORRESPOND WITH DANIEL'S (AS WELL AS JOHN'S) PROPHECY?***

EVENTS OF THE LAST GENERATION OF THE JEWISH NATION WERE, ASIDE FROM THE BIRTH OF JESUS, THE FOLLOWING:

1. **THE DAILY TAKEN AWAY AND ABOMINATION OF DESOLATION SET UP-** THE REJECTION OF JESUS AND JESUS' DOCTRINE BY THE APOSTATE JEWISH CHURCH AND HIS CRUCIFIXION (A.D.31) (Daniel 8:10-12), with Caesar and Roman Sun God SPIRITUALLY SET UP in the place of the daily.

2. **ARRIVAL AND STANDING OF ABOMINATION OF DESOLATION (ROMAN BEAST)** (A.D. 66) to claim worship in the place of the God of Israel. The "GOD" of Rome against the God of Israel.

3. **DESTRUCTION OF JERUSALEM** AND THE NATION 3.5 YEARS LATER. (A.D. 70)

Did Daniel also see a parallel or corresponding sequence of events for the last generation of the world? Yes, he did, and we read this in Daniel 12. Teachers who teach that Daniel 12 prophetic periods are historically fulfilled and past should beware. They should study Daniel 12 again because nothing whatsoever indicate that those periods are in the past. The pioneers made the mistake because they believed, SINCERELY, that Jesus was coming in 1844 A.D. But we today have no excuse for teaching this error.

THE APOCALYPTIC EVENTS OF THE LAST GENERATION OF THE WORLD

1. DAILY TAKEN AWAY- JESUS AND JESUS' DOCTRINE REJECTED BY APOSTATE PROTESTANT CHURCHES JUST AS THE JEWISH CHURCH DID- RESULTING IN

LEGISLATION OF SUNDAY LAW AND OTHER PAGAN DOCTRINES OF THE ROME. By this, they crucify Jesus spiritually.

2.ABOMINATION OF DESOLATION-THE SAME ROMAN BEAST (AS PAPAL ROME) IN ITS THIRD COMING-FIRST AS ANCIENT ROMAN BEAST, SECOND AS MEDIEVAL PAPAL ROME, AND THIS THIRD AND LAST AS APOCALYPTIC PAPAL ROMAN BEAST (BEAST FROM THE BOTTOMLESS PIT-REVELATION 17).

3.DESTRUCTION OF APOSTATE RELIGIONS BY THE BEAST. THEN THE BEAST AND ITS WORLD'S DESTRUCTION BY JESUS.

THE DATES OF THE APOCALYPTIC EVENTS OF THE LAST GENERATION OF THE WORLD

1. A.D. 2031(1290 DAYS PRECEDING THE COMING OF ABOMINATION OF DESOLATION)- DAILY TAKEN AWAY (LEGISLATION OF SUNDAY LAW AND OTHER PAGAN DOCTRINES OF ROME. POPE DECREED BY AMERICA TO BE THE HEAD OF ALL RELIGION. PERSECUTION BEGINS FOR THE TRUE CHURCH OF GOD.

2. A.D. 2034 - THE 144000 SEALED AND CLOSE OF PROBATION. ABOMINATION OF DESOLATION ASCENDS TO POWER FROM THE "BOTTOMLESS PIT"(FULL AND COMPLETE REVIVAL OF IMPERIAL ROMAN BEAST) TO RULE FOR 1260 DAYS. THEN THE ANCIENT OF DAYS SITS AND SEVEN TRUMPET JUDGEMENTS BEGIN.

3. A.D. 2038 - JESUS(MICHAEL) STANDS UP. SEVEN LAST PLAGUES POURED OUT. BABYLON THE GREAT DESTROYED, JUST AS THE APOSTATE JEWISH CHURCH AND NATION WAS DESTROYED. THEN JESUS COMES IN A DAY AND HOUR THAT IS NOT REVEALED.

ADDRESSING THE FEAR OVER KNOWING THE TIMES

Unfortunately, Seventh-day Adventist church has not as yet realized that God has given her the complete timeline of the Great Controversy, up to the year of Jesus's coming, withholding only the DAY and HOUR. The impediment to this knowledge is to be traced to the same factor that led to the Great Disappointment of October 22, 1844. The same problem and mistake William Miller made is still with present day Seventh-day Adventists. Seventh-day Adventists almost unanimously believe William Miller's only mistake is the misidentification of the sanctuary that is to be cleansed. William's mistake was two-fold however, namely, misunderstanding of what the SANCTUARY in Daniel 8 means and secondly misinterpretation of Daniel 12 prophetic periods, among many other prophecies of Daniel and revelation.

It is important to know the scope of William Miller's work. William Miller had done excellent study on the prophetic periods relating to the last generation of the Jewish nation. From his study we have the events of the Last generation of the Jewish nation with their dates as follows:

- Jesus's baptism-27 A.D.
- Jesus's death-31 A.D.
- Close of probation for the nation-34 A.D.
- From the year of Jesus's baptism, and applying Luke 3:23, we have Jesus's year of birth as 2 B.C., which also marked the beginning of that last generation of the Jewish nation-church.

But with regards to the last generation of the world, which is found in Daniel 12, William Miller made a mistake that stems from his belief that the sanctuary to be cleansed is the earth, and that the last generation of the earth will end on October 22, 1844.

Consequently he interpreted Daniel 12 prophetic periods as sub-sections of the 2300 DAYS(YEARS) of Daniel 8:14. It was an innocent mistake, but a revelation of the danger of not examining traditional teachings against scriptures, of not being certain that what one believes is actually and strictly based on scriptures. Just accepting the then common belief and teaching that the earth is the sanctuary caused the movement to suffer a Great Disappointment, comparable only to the one the disciples experienced when their Lord and savior was crucified.

WILLIAM MILLER AND A POST-DISAPPOINTMENT ANALYSIS OF DANIEL 12-WHY HE REJECTED ANY FURTHER DATE SETTING

William Miller well understood that Daniel 12 prophetic periods are about the LAST PERIODS OF THE WORLD AND GREAT CONTROVERSY, which he mistakenly thought will end on October 22, 1844. But after the Great Disappointment, he could not continue to believe that Daniel 12 prophetic periods are subsets of the 2300 DAYS. Since those prophetic periods plainly and clearly lead down to the end of the Great Controversy, that is when Jesus will come, William Miller would have come to the conclusion that they cannot be part of the 2300 DAYS, and must therefore be periods beyond the 2300 DAYS. With there not being any prophetic period from scriptures to **LINK THE 2300 DAYS AND DANIEL 12 PROPHETIC PERIODS**, William Miller must have come to the conclusion that there is no way to determine exactly when Jesus will come. This is why he never again accepted any day and hour setting.

THE MISSING LINK GIVEN TO ELLEN G. WHITE AND WHY SHE DID NOT FULLY UNDERSTAND IT: **THE ACTUAL PURPOSE OF THE 4000, 6000 YEARS MESSAGES**

Even after realizing the important reason why October 22, 1844, was not the day and hour of Jesus' coming, Ellen G. White and other pioneers continued to hang on to most of the premises that led to that wrong conclusion. They still continued to believe that Jesus will come in their generation. Because of this, they continued to sustain all the false interpretations of the books of Daniel and Revelation. Because of this Ellen G. White could not even fully understand the purpose and accurate interpretation of her 4000, 6000 years prophetic messages.

To understand the 4000, 6000 years prophetic message requires accurate knowledge and understanding of Daniel 8-12. The first part, which is the 2300 DAYS part was understood by Daniel, as he indicated in Daniel 10:1. But when he received the second vision, Daniel 11-12, in which the angels introduced three other prophetic periods-1260, 1290, and 1335 DAYS, he was stomped. Thoroughly confused and desperate and frustrated, he made his confusion known, "And I heard, but I understood not:", then asked in desperation and frustration,: O my Lord, what shall be the end of these things?"

What Daniel needed at that time was for the Angel to tell him how many years will connect from the end of the 2300 DAYS to the three prophetic days. But even if he got this knowledge, he would still have needed the angel to tell him how the three prophetic periods are to be arranged relative to each other. But even if the Angel had also done so, and went no further, Daniel would still not have known the day and hour of his people's deliverance because, as will be seen from accurate arrangement of the three prophetic periods, there is an unknown, indeterminate period between the first two prophetic periods and the third one, the 1335 DAYS, during which the deliverance will occur.

There was a whole lot of information that Daniel needed to be given for him to receive that precious knowledge and information that he so desperately desired, namely, THE EXACT TIME OF HIS PEOPLE'S DELIVERANCE AND THE PERMANENT DEFEAT AND DESTRUCTION OF THAT ENEMY POWER THAT HAS RELENTLESS AND PROTRACTEDLY OVERCOME HIS PEOPLE. But the angel lovingly informed him that those knowledge he sought are sealed, to be opened only to the generation that would need it and at the time meant for it, which is TIME OF THE END.

So, the information Daniel sought for, which the angel told him that they are sealed are threefold:

1. The period of time linking the 2300 DAYS of Daniel 8:14 with Daniel 12 three prophetic periods.

2. How the three prophetic periods are to be arranged relative to each other.

3. The exact time the Deliverance will occur, starting with when Michael will stand up, and shortly after when Michael (Jesus) will come and the resurrection and immortalizing of saints accomplished.

ELLEN G. WHITE RECEIVED THE FIRST OF THE ABOVE THREE LISTS OF KNOWLEDGE, which is what the 4000, 6000 years messages were about. With those messages God gave her the link between the 2300 DAYS of Daniel 8:14 and Daniel 12 prophetic periods.

The second in the list requires that Daniel 12 be understood as future events that have not yet been fulfilled. She never had this knowledge because she, like the other pioneers, were still ensnared by William Miller's mistake of subsuming all prophetic periods under the 2300 DAYS. Thus all prophetic periods including Daniel 12 prophetic periods are presumed to have ended on October 22, 1844.

So, still believing the mistaken idea that the Daniel 12 prophetic periods are years, and subsumed in the 2300 days(years), she could not see the actual purpose for her 4000, 6000 years messages, which is to provide the 2000 YEARS LINK BETWEEN THE GREAT CONTROVERSY EVENTS OF THE LAST GENERATION OF THE JEWISH NATION AND THOSE OF THE LAST GENERATION OF THE WORLD.

Lastly, when the three prophetic periods are accurately arranged, it will then be discovered that because of the gap of unknown period of time between the second prophetic period, the 1260 days and the third, the 1335 days, the day and hour of Jesus's coming can never be known from any prophetic period. It has to be declared by God Himself extra-canonically. But until the DAY and HOUR knowledge is given, Seventh-day Adventist church is blessed with the knowledge of THE YEAR OF OUR LORD'S COMING.

DANIEL 12 PROPHETIC PERIODS: ARE THESE LITERAL AND FUTURE OR SYMBOLIC OF YEARS ALREADY FULFILLED?

Daniel 12 has three prophetic periods, and these are 1260, 1290, and 1335 DAYS. Of the three only the 1260 days is mentioned more than once in the books of Daniel and Revelation. The other places the 1260 DAYS are mentioned are Daniel 7, Revelation Chapters 11, 12, and 13. However if we understand the events that are to occur in the other prophetic periods, 1290 and 1335 days, we will realize that even though they are not expressly stated again, they are alluded to through the events that occur in them. We will begin making our case for the literal and future fulfillment of the prophetic days by examining Daniel 12, where the three are mentioned together alongside each other.

THE QUESTION AND THE ANSWERS: It all begins with the quintessential question asked by one of two heavenly personalities. Understanding this question is so significant that once understood, there is no logical manner that the prophetic periods could ever be seen as other than literal and future. The passage starts as follows:

"Then I, Daniel looked, and, behold, there stood other two, the one on this side of the bank of the river, and the other on that side of the bank of the river. [6] And one said to the man clothed in linen, which was upon the waters of the river, ***How long shall it be to the end of these wonders?"***-Daniel 12:5,6.

It is in response to these question that the three prophetic periods are given for Daniel and the church. Therefore a careful, thorough, and precise distillation of this question is very crucial, as they inform what the answers mean and inform.

NARROWING DOWN THE QUESTIONS

What actually was the heavenly being asking, and what did Daniel ask? When the two questions (which are in furtherance of the same question in Daniel 8:14) are narrowed down to specifics, what do we come up with as the nitty gritty of the question? In order to narrow the question down to the basics, we need answer another question, namely: ***WHAT DOES THE ANGEL MEAN BY "THESE WONDERS", AND WHICH IS THE LAST(END) OF THE WONDERS?***

Proceeding without answering these questions is absurd, and will only lead to a gross misunderstanding and misinterpretation of Daniel 12. The context of the above questions is the events of the vision that began in Daniel 11 and ended in Daniel 12:1-3. It is these events that the Angel is referring to as ***THESE WONDERS.*** The last one, as will be clear to the reader, therefore, is the event or wonder of the TWO RESURRECTIONS and the IMMORTALIZING OF THE saints ("Daniel's people"):

Daniel 12:1-13 KJVS

And at that time shall Michael stand up, the great prince which standeth a for the children of thy people: and there shall be a time of trouble, such as never was since there was a nation even to that same time: and at that time thy people shall be delivered, every one that shall be found written in the book. [2] And many of them that sleep in the dust of the earth shall awake, some to everlasting life, and some to shame and everlasting contempt. [3] And they that be wise shall shine as the brightness of the firmament; and they that turn many to righteousness as the stars for ever and ever".

Armed with this knowledge, we can now accurately rephrase the question as follows:

1. **How long shall it be to the standing up of Michael and the resurrection and immortalizing of all saints?**

2. **How will Michael end all these suffering of the sanctuary and the host(saints)?**

3. **When will Jesus (as Michael) stand up, and come to deliver the saints, resurrect, and immortalize the saints?**

4. **In short: <u>WHAT IS THE DAY AND HOUR OF JESUS' COMING</u>?**

It is therefore to the question of the day and hour of Jesus' second coming, that the other heavenly person gave the answers comprising the three prophetic periods, 1260, 1290, and 1335 days. To a sincere and intelligent reader, it becomes apparent why the three prophetic periods can never be past fulfilled because they are answers to question about a future event, the second coming of Jesus Christ.

Now that we have clear understanding of what the question is, let us now look at the answers given, which are as follows:

"And I heard the man clothed in linen, which was upon the waters of the river, when he held up his right hand and his left hand unto heaven, and sware by him that liveth for ever that it shall be for a time, times, and an half; and when he shall have accomplished to scatter the power of the holy people, **all these things shall be finished**".

Whereas Angel Gabriel had explained the 2300 DAYS timeline and events to Daniel earlier on and he indicated that he understood the matter (See Daniel 10:1), her, Daniel clearly manifests lack of understanding, meaning that the prophetic period that he just heard, the 1260 days, is not part of the 2300 DAYS.

"And I heard, but I understood not: then said I, O my Lord, what shall be the end of these things?"

As a clear evidence that the matter of the 1260 days concerns something Daniel cannot and must not know, the Angel told him:

"Go thy way, Daniel: for the words are closed up and sealed till the time of the end. [10] Many shall be purified, and made white, and tried; but the wicked shall do wickedly: and none of the wicked shall understand; but the wise shall understand".

As though Daniel's was not already confused enough, the angel adds to his confusion with the other two prophetic periods.

“And from the time that the daily sacrifice shall be taken away, and the abomination that maketh desolate set up, there shall be a thousand two hundred and ninety days. [12] Blessed is he that waiteth, and cometh to the thousand three hundred and five and thirty days. [13] But go thou thy way till the end be: for thou shalt rest, and stand j in thy lot at the end of the days”.

So, evidently, Daniel did not know what to make of the three answers the heavenly person had rattled out. In his quest to understand, the angel told him to shut it off, and leave till the time of the end, when the answer will come. These prophetic periods can therefore in no way be part of the 2300 DAYS, for which Angel Gabriel had already given him sufficient answers and explanation. For instance, Daniel knew when the 2300 DAYS will begin and therefore when it will end. He also knew that it was not 2300 literal days but 2300 years because he must have figured that 2300 literal days (approximately 6 years plus) would not be enough for his people to rebuild Jerusalem and the sanctuary to have it cleansed. Therefore he knew that the 2300 days must be 2300 years for which only 390 years is given to his natural people and nation.

So, the fact that Daniel never understood the three prophetic days of Daniel 12 is a clear evidence that they are not part of or subsumed in the larger 2300 years. Couple this fact with the other, namely, that the question to which the three prophetic periods were given to answer is, when distilled down, about the DAY AND HOUR OF DELIVERANCE OF THE SAINTS BY MICHAEL(JESUS) AND RESURRECTION OF DEAD SAINTS.

Therefore the three prophetic periods must be literal days that are still future, as the question of the day and hour of Jesus’ coming is future. Otherwise we will be looking at 1260, 1290, and 1335 years farther for the second coming of Jesus.

WHO CAN POSSIBLY LIVE AND ENDURE PERSECUTION FOR 1290 + 1260 YEARS AND THEN WAIT TO COME TO 1335 YEARS PERIOD FOR THEIR DELIVERANCE? THE LAW OF DEATH CANNOT PERMIT THIS

If, as Historical interpreters of Daniel 12 prophetic periods presume, these periods are YEARS, then God is guilty of asking that which He has made impossible with the Law of death and mortality of sinners.

It would mean that God is asking people who would have been enduring persecution for 2550 YEARS (1290 + 1260), to still wait patiently and in faith in order to come to the 1335 YEARS period of blessing.

Any astute bible student will see the impracticability and impossibility of the implication from regarding these prophetic periods as YEARS.

Lastly, consider the last statement the Angel made to Daniel:

Daniel 12:13 KJVS

“But go thou thy way TILL THE END BE: for thou SHALT REST (IN DEATH), and stand in thy lot AT THE END OF THE DAYS”.

There is no denying what DAYS the Angel is talking about here, for the context is immediate. It is the three prophetic periods that the Angel is referring to, clearly, and especially the longest one which is associated with the blessing of the resurrections and immortality of the saints, the 1335 DAYS.

The expression, STAND IN THY Lot will of course resonate with Daniel for it connotes how his ancient people inherited Canaan and will re-inherit the land after the end of Babylonian captivity. But Daniel is not intellectually or spiritually challenged and dull, and would quickly realize that it is not earthly inheritance that the Angel is talking about here.

He will realize that the only way for him to receive this inheritance after he has died is to experience resurrection, and that God will not pull him out of his rest in death just to give him an earthly, perishable plot of land, or even all the money in this world. Therefore Daniel knew the Angel was talking about the ETERNAL, IMMORTAL GLORY that will be the inheritance of all the saints when THE END OF THE CONTROVERSY comes.

Since this is what the Angel is talking about, and since the Angel clearly says that this will occur at the END OF THESE PROPHETIC DAYS, with the 1335 DAYS being the one associated with this blessing, we have clear evidence therefore, that this 1335 DAYS, and the other two by association, are future and not 1335 YEARS that Historical teachers claim as ending in 1843 A.D.

Did Daniel stand in his lot in 1843 or 1844? Clearly not. Uriah Smith did his best literary calisthenics to explain the expression STAND IN THY LOT as being the INVESTIGATIVE JUDGMENT DECISION made on the case of Daniel beginning in 1844 A.D. Well, that's a nice try, but it just doesn't pan out to even the least literary endowed person.

Thus far, we have exhaustively proved, from the context of Daniel 12, that the 1260 DAYS and the associated prophetic periods, 1290 and 1335 DAYS, are literal days and are still to be fulfilled in future.

But there are FIVE other mentions of the 1260 DAYS prophetic period. Do these other places mentioned suggest that it is 1260 YEARS and already fulfilled? Well, let's go to Daniel 7 next.

THE 1260 DAYS IN DANIEL 7

Daniel 7 traces the history of the FOUR WORLD EMPIRES- BABYLON, MEDIA PERSIA, GREECE, and ROME. But let's cut down to the fourth, ROME, because it is during the period allotted to this empire that the 1260 DAYS PERSECUTION occurs.

How it unfolds is that in the latter days of Rome, it was divided into TEN SUB'S KINGDOMS represented as TEN HORNS. But in the midst of these ten horns, a horn emerges that starts out as a small horn. Thus for a brief period, there were 11 HORNS on the ANCIENT ROMAN BEAST.

Eventually, this eleventh, small horn gobbled up THREE HORNS, and the entire beast ended up with EIGHT HORNS, with the small horn becoming FAT, from eating up the three horns that stood its way to dominance. The fat stature of course represents the fact that it expanded its power and increased its jurisdiction with those of the three it defeated.

It is this horn that started off little and later became big, that the 1260 DAYS are mentioned in connection to. As Daniel narrated:

Daniel 7:20-22 KJVS

"And of the ten horns that were in his head, and of the other which came up, and before whom three fell; even of that horn that had eyes, and a mouth that spake very great things, whose look was more stout than his fellows. [21] I beheld, and the same horn made war with the saints, and prevailed against them; [22] UNTIL the Ancient of days came, and judgment was given to the saints of the most High; and the time came that the saints possessed the kingdom".

Here is a key understanding a reader should fasten onto quickly, and that is that A PERIOD OF TIME IS GIVEN TO THIS HORN POWER during which it was to exist and operate to persecute God's people. Daniel sees this allotted period come to an end with the COMING AND SITTING OF THE ANCIENT OF DAYS IN JUDGEMENT.

This point is very significant, because it means that beyond the point the Ancient of days sits, this horn power has no more authority and capability to OVERCOME or PREVAIL AGAINST GOD'S PEOPLE. In other words, when the Ancient of Days sits, this power would have come to the end of its dominance. It will never again emerge, but instead soon after it becomes destroyed.

The need to make this point is because, Uriah Smith and all other Historical teachers of his day, had assumed that the time when thus power met its FINAL DEMISE AND NEVER TO RISE AGAIN, is 1798 A.D. It is on this false premise that they placed the time for the coming of the Ancient of days at OCTOBER 22, 1844, A.D. These historical teachers never saw that this power will arise again, because they believed that Jesus would come in their lifetime, which leaves no room in time for this horn power to re-emerge and achieve second world dominance and persecute God's people again.

Why is this important? Because we are about to read a parallel statement by Daniel where he mentions the 1260 DAYS as a period that leads up to the Ancient of days sitting. When we line up the two parallel statements, it becomes obvious that the 1260 DAYS is the LAST BATCH OF TIME JUST BEFORE ANCIENT OF DAYS SITS, which means that after the 1260 DAYS, this horn power is never to rise up again.

Here are the two parallel passages side by side.

“beheld, and the same horn made war with the saints, and prevailed against them; [22] UNTIL THE ANCIENT OF DAYS CAME, and judgment was given to the saints of the most High; and the time came that the saints possessed the kingdom”-Daniel 7:21-22 KJVS

And he shall speak great ab words against the most High, and shall wear out the saints of the most High, and think to change times and laws: and they shall be given into his hand UNTIL A TIME AND TIMES AND THE DIVIDING OF TIME. [26] But the judgment shall sit (ANCIENT OF DAYS SHALL SIT), and they shall take away his dominion, to consume and to destroy it unto ah the end-Daniel 7:25-26 KJVS

Do you notice the emphasized phrases? They indicate that the two passages are parallel, meaning they are saying the same thing, each giving complementary information that provide additional information. It is called PARALLELISM.

So, the two passages are saying that the 1260 DAYS is the LAST BATCH OF TIME given to this power after or beyond which it will never arise again.

The question, therefore, is this: Did this happen in 1798 A D. Was 1798 A.D. the last time for the existence of this beast? The answer is no, because we see that in Revelation 13 and 17, that this beast is to have TWO LIVES AND EXISTENCES. It is soon to regain world dominance, which Uriah Smith and co-historicists could not have envisaged.

William Miller Advent theologians, which included the pioneers of SEVENTH-DAY ADVENTIST CHURCH, built a whole lot of doctrines on the false premise that Jesus was coming on October 22, 1844, or at the latest during the generation. SEVENTH DAY ADVENTISTS will therefore need to UNLEARN A WHOLE LOT OF THEOLOGY AND LEARN NEW ONES.

It is on this note that Ellen G. White was shown in vision that A TEACHER, or at least FEW TEACHERS will arise in SEVENTH-DAY ADVENTISTS CHURCH, who will PREACH AND TEACH STRAIGHT TRUTHS THAT WILL BRING ABOUT SHAKING IN THE CHURCH DUE TO MANY REJECTING THESE TRUTHS.

The preachers of these truths will be labeled by the General Conference as fanatics and false teachers, and the movement that God will begin will be denounced and dismissed as an OFFSHOOT. But the power of God will attend these men and women and children, and they will finish the work rapidly.

THE 1260 DAYS IN REVELATION 11

In Revelation 11, we have what can be considered the most logical, irrefutable evidence that the 1260 DAYS are literal days that are to be fulfilled in the future. Revelation 11 ranks the same with Daniel 12 in terms of how logically it shows that the 1260 DAYS can never be YEARS already fulfilled. We will approach the evidence from two perspectives, and therefore under two sub-sections.

THE TWO WITNESSES AND THE BEAST THAT WILL KILL THEM EXACTLY AT THE END OF 1260 DAYS

Revelation 11:3-7 KJVS

"And I will give power unto my two witnesses, and they shall prophesy a thousand two hundred and threescore days, clothed in sackcloth...

And when they shall have finished their testimony, the beast that ascendeth out of the bottomless pit shall make war against them, and shall overcome them, and kill them".

Notice that the text does not say the beast that ascended or rose up **FROM THE SEA**, a clear evidence that it is not referring to the Medieval Papacy, which rose up from the sea (See Revelation 1:1)

Rather it is talking about a later time when it will arise a second time. We know that even though this beast's rise is not from the sea, it is still the same Papal Beast, which means that the Papal beast will have two phases of existence during its lifetime. This is evident from the following activities that John was shown:

Revelation 13:3 KJVS

"And I saw one of his heads as it were **WOUNDED TO DEATH** (First phase existence ends); **AND HIS DEADLY WOUND WAS HEALED** (Second phase existence commences):

and all the world **wondered after the beast**" (Notice when the whole world wondered after the beast, which is after the deadly wound is healed. This is an important point we will develop when we discuss this 1260 DAYS in Revelation 13)

Again in Revelation 17, John also received another information that corroborates the fact that this beast has two phases of existences, and on this he wrote:

Revelation 17:8 KJVS

"The beast that thou sawest **WAS**, and **IS NOT** (DIED FROM DEADLY WOUND); and **SHALL ASCEND OUT OF THE BOTTOMLESS PIT** (AT A LATER DATE IN FUTURE), and GO INTO PERDITION: and they that dwell on the earth ***shall wonder***, whose names were not written in the book of life from the foundation of the world, when they behold the beast that was, and is not, and yet is".

Yet again, on this same note, John wrote:

Revelation 17:11 KJVS

"And the beast that **WAS**, and **IS NOT**, even **HE IS THE EIGHTH**, and **IS OF THE SEVEN**, and goeth into perdition".

When it arises this second time, it will be, not from the sea as was the case with its Medieval rise, but rather ***out of the bottomless pit***, which is clearly at a future time, when the deadly wound is fully healed. It will then go into perdition shortly afterwards, indicating a very short second existence, in significant contrast with its first existence that was for over a thousand years. We know that MEDIEVAL PAPACY did not go into perdition yet, but rather was only dead momentarily.

Notice also that in its second existence, it will be the **EIGHTH HEAD**, meaning it will have the functional head that is the eighth. Since the beast was not originally seen by John with eight heads, it means that ***one of the fallen heads*** will resurrect or rise up, after the seventh head. This is what John saw happen and what he means by that the eight head is ***"of the seven".*** It means that this eight head is one of the original heads that had fallen or died.

So when this beast ascends from the bottomless pit, it will be the **"LAST OF THE LAST" kingdom,** the very beast government that Jesus will destroy when he comes. The head of this second and last PAPAL BEAST government must therefore be the one Paul called THE MAN OF SIN in

2 Thessalonians 2:8 KJVS

"And then shall that Wicked be revealed, whom the Lord shall consume with the spirit of his mouth, and shall destroy with the brightness of his coming:"

When the deadly wound is fully healed, it is then that it will have risen from the bottomless pit, and this is clearly in the future still. Only a deep-seated bias can prevent anyone from seeing the merit of these reasonings.

Based on this accurate knowledge therefore, and knowing that it will kill the Two witnesses exactly at the end of the 1260 DAYS, we have here, therefore, an incontrovertible proof that the 1260 DAYS is still in the future. As at 1798 A.D., PAPAL BEAST has clearly not risen from the bottomless pit, not to talk of 533 (or 538 A.D., as historical interpretation supposes as when Medieval Papacy began its rule). The event of this beast killing the two witnesses after 1260 days of their witnessing, cannot, in any sense, preceded the rise of this beast from the bottomless pit. The conclusion is inescapable, namely, the 1260 DAYS and the work of the two witnesses are future, which makes the 1260 DAYS literal days, since the alternative, a future 1260 YEARS is untenable.

THE CORRESPONDING PROOF THAT THE SEVEN TRUMPETS ARE STILL FUTURE:The 1260 DAYS are intertwined with the FIRST SIX TRUMPETS. Notice that the SIXTH TRUMPET comes to an end **3.5 DAYS after the end of the 1260 DAYS:**

Revelation 11:11-14 KJVS

"And after <u>THREE DAYS AND AN HALF</u> the Spirit of life from God entered into them, and they stood upon their feet; and great fear fell upon them which saw them. [12] And they heard a great voice from heaven saying unto them, Come up hither. And they ascended up to heaven in a cloud; and their enemies beheld them. [13] <u>AND THE SAME HOUR</u> was there a great earthquake, and the tenth part of the city fell, and in the earthquake were slain of men seven thousand: and the remnant were affrighted, and gave glory to the God of heaven. [14] <u>THE SECOND WOE (SIXTH TRUMPET)</u> is past; and, behold, the third woe cometh quickly".

So, as you can see, the sixth trumpet did not end in Revelation 9:21, as we have been wrongly taught for decades by historical interpreters of the 1260 DAYS prophecies.

Since the sixth trumpet is **1 YEAR, 1 MONTH 1 DAY, 1 HOUR** long, and the fifth trumpet is **5 MONTHS** long, we can count back approximately 1 YEAR 5 MONTHS from the 1264TH DAY when the sixth trumpet ended, to when the fourth trumpet ends, which will be at approximately the **749TH DAY of the 1260 DAYS**. This leaves on about 749 days for the first four trumpets. **<u>Therefore we can confidently say that the FIRST SIX TRUMPETS span the 1260 DAYS.</u>**

Since we have proved irrefutably and uncontestably, that the 1260 DAYS are in the future, we equally have irrefutable and incontestable proof that the SEVEN TRUMPETS are still in future, contrary to the historical school of thought and its teachers, which have always seen the first six trumpets as having been fulfilled in the past.

The spirit of prophecy warned the church against complacency regarding the study and understanding of Daniel and Revelation. Unfortunately, our teachers and ministers are exhibiting the very LAODICEAN ATTITUDE in regard to Daniel and Revelation, believing they have no need of anything, rich in goods, not knowing how wretched, and poor, and blind the church is in this matter. With regards to Daniel and Revelation, the church indeed has much to unlearn and so much to learn. May God help His church in Jesus' name Amen.

THE TWO WITNESSES AND THE CONCLUSION OF REVELATION 10 COMMANDMENT: "THOU MUST PROPHESY AGAIN"

REVELATION 10 "TIME NO LONGER" STATEMENT EXPLAINED

In this second perspective, we examine the Revelation 1260 DAYS from the lens of its connection with Revelation 10 commission, "THOU MJUST PROPHESY AGAIN. We will notice that it is at the point where the SIXTH TRUMPET is about to conclude that Revelation 10 vision is given to John, clearly as a PARENTHETICAL INTERLUDE. The very first scene was of the SEVEN THUNDERS, the LITTLE OPEN BOOK, and the declaration under oath to Jehovah: THERE SHALL BE TIME NO LONGER". The natural question is: TIME NO LONGER FOR WHAT? What is the event for which its appointed time is declared by this Angel to have ended?

The clue comes from what has been going on until **Revelation 10(and Revelation 11:1-14)** vision is introduced as a PARENTHETICAL INTERLUDE. John had been seeing the SIXTH TRUMPET JUDGMENT, and just as it was about to conclude, the Angel introduces Revelation 10 and Revelation 11:1-11 in parenthesis. It is important to notice that baring the chapter distinction, Revelation 10 and Revelation 11 constitute ONE

VISION, and this is how one can understand the message God intends for the church. Chapters are not original to scriptures. They are introduced by translators. They are sometimes helpful in, but sometimes they are not. Revelation 10 naturally flows into and continues into Revelation 11, as one vision, and up to verse 14 of chapter 11, we have a parenthetical vision that began just before the SIXTH TRUMPET officially ended.

WHAT IS THE PURPOSE OF PARENTHESIS IN A NARRATIVE? IS IT MERELY AN INTERRRUPTION OR AN ENHANCEMENT OF THE NARRATIVE?

A parenthesis circles back to the point that was reached in the original narrative before it was introduced. It gives additional information that always terminates at the point at which the parenthesis was introduced. Thus, you will notice that Revelation 10 is saying essentially, THOU MUST PROPHESY AGAIN TILL THE END OF THE SIXTH TRUMPET. Notice also how Revelation 11:1-14 corroborated this also by circling back to the end of the SIXTH TRUMPET. **So, the END OF THE SIXTH TRUMPET provides the IMMEDIATE CONTEXT for the parenthetical messages of Revelation 10 and Revelation 11:1-14.**

FURTHER EXPLANATION ON REVELATION 10: THE CONTEXTS OF THE DECLARATION: "TIME NO LONGER".

With the commandment and commission to prophesy again, in Revelation 10, must also come the revelation of when this mandate will end, since the saints will not preach all the way to heaven. There must therefore be an allotted time given to accomplish this commission. Did prophecy indicate when the commission to prophesy again is to end? Yes.

When we discount the chapter demarcation between Chapter 10 and 11, we discover that the last verse of chapter 10 natural continues into verse 1 of Chapter 11. What this means is that the commission to prophesy again, which had begun on October 22, 1844, continue into the work of the Two Witnesses. The work of the Two Witnesses is therefore part of the later and last phase of the commission to PROPHESY AGAIN, which clearly puts the work of the 1260 DAYS work of the Two Witnesses future to October 22, 1844.

With regards to the Last Generation of the Jewish nation-church, the preaching mandate and commission ended on the day the Abomination of desolation stood in the holy place, in 66 C.E., in accordance with Jesus' DUAL PROPHECY. All the Christian disciples fled in faith and obedience to Jesus' prophecy. The time allotted to warn the nation and people of the coming destruction was over on that day in 66 C.E. Jesus' DUAL application prophecy (which relates to Daniel 8, 9, and 12 prophetic periods, and Revelation 12, also point to the Last Generation of the world, and also give the same revelation regarding when the gospel PREACHING("PROPHESYING") mandate and commission is to cease.

Likewise, in the last generation of the world, the clue to knowing when to stop preaching the gospel message remains whenever the Abomination of desolation will stand in the holy place. This is the point at which the "virtuous woman" (the church) is given wings of an eagle to flee into the wilderness where she is to be nourished for 1260 DAYS (Revelation 12). The Abomination of desolation will stand at end of the 1290 DAYS, which marks the beginning of the 1260 DAYS, and at this time all the saints **except the Two Witnesses** are to cease preaching and flee into the Wilderness. But in uniqueness to the Last generation of the world, God gives the Two Witnesses power to continue and conclude the commission begun on October 22, 1844. They are to engage in the last phase of the work of witnessing(prophesying), for the next 1260 DAYS, while the Woman is in the wilderness. Why is this 1260 DAYS so important, and what will they be prophesying about, and what is the goal?

THE UNIQUE COMMISION OF THE TWO WITNESSES

God chose the Two witnesses to continue the work of prophesying begun on October 22, 1844. After the church has escaped into the wilderness, the Two Witnesses are to continue the commission and mandate. The two are said to prophesy IN SACKCLOTH. This is different from the first part of the commission the larger Church was commanded to undertake from October 22, 1844, to do. Recall that the Abomination of desolation is said to CONTINUE FOR FORTY-TWO MONTHS (See Revelation 13), and that it is called the GENTILE POWER.

The purpose and content of the work of the two witnesses is to PROPHESY AGAINST THE GENTILE POWER (ROME-PAPAL AND HER ADHERENTS) THAT WILL BE TREADING UNDERFOOT THE OUTER COURT, THE HOLY CITY. They are not prophesying salvation because the SEALING OF THE LAST REMAINING SAINTS ALIVE (REMNANT OF THE SEED OF THE WOMAN), that is the 144000, is already completed. Rather they are PROPHESYING GOD'S JUDGMENT AGAINST THE WICKED "GENTILE" POWER (THE BEAST AND HIS GOVERNMENT AND THE WICKED WHO ARE WORSHIPING HIM), which is the Abomination of desolation power.

The two Witnesses are calling down the SEVEN TRUMPET PLAGUES, just as Moses called down judgment on Pharaoh and Egypt, and just as Elijah called down judgment on Ahab, Jezebel and the nation that had largely followed Baal. Their message is not of the Grace and Mercy of God but instead of the JUDGMENT OF GOD upon the wicked power and forces of evil working against God and His purposes and persecuting and killing His people, hence the sackcloth.

So, contrary to the false evangelical-protestant theology that presumes that during the last 1260 DAYS tribulation, the Jewish nation will be redeemed, and that the 144000 are exclusively natural Jews, Revelation 11 prophecy is teaching instead that the Jewish nation will double down on their rejection of Jesus, the True Messiah and King, and will accept and worship the False Messiah, the ANTICHRIST or MAN OF SIN. It is against this sin, of validating the False Christ against God, the same kind of sin that led to her destruction in A.D.70, that the Two Witnesses are to prophesy against.

This 1260 DAYS work of the Two Witnesses concludes the commission given in October 22, 1844, as shown to John in Revelation 10 vision: "Thou must prophesy again". At exactly the day and hour the 1260 days end, the Two Witnesses are killed by THE BEAST THAT WILL ASCEND FROM THE BOTTOMLESS PIT. Their resurrection and ascension 3.5 DAYS later, coincides with the END OF THE SIXTH TRUMPET JUDGMENT.

It is at this point that the declaration by the Angel that "THERE SHOULD BE TIME NO LONGER" (Revelation 10:1), is fulfilled, which means TIME NO LONGER FOR THE COMMISSION TO PROPHESY AGAIN. This is the time when the following declarations will be fulfilled:

Revelation 11:14 KJVS

"The second woe is past; and, behold, the THIRD WOE COMETH QUICKLY".

Revelation 22:11-12 KJVS

"He that is unjust, let him be unjust still: and he which is filthy, let him be filthy still: and he that is righteous, let him be righteous still: and he that is holy, let him be holy still. [12] AND, behold, I COME QUICKLY; and my reward is with me, to give every man according as his work shall be".

It is at this time, when the SIXTH trumpet concludes, which is at the end of the 1260 DAYS (3.5 DAYS after) that the declaration of TIME NO LONGER fulfills. PROBATION IS ENDED FOR THE WORLD at this point, and shortly after, the WRATH OF GOD will be poured out with the SOUNDING OF THE SEVENTH TRUMPET.

Therefore the declaration THERE SHOULD BE TIME NO LONGER is not without any context, and the context is WHEN THE SIXTH TRUMPET ENDS, AT THE END OF THE 1260 DAYS, and it signifies that PROBATION HAS CLOSED.

THE 1260 DAYS IN REVELATION 13

We will not go into how it is determined that Revelation 13's first beast is Papal Roman beast. The reader is encouraged to study Adventist books that make this fact abundantly clear. The main focus here will be to prove that the 1260 DAYS apply, not to the Medieval phase of the beast, but instead to the later and still future phase of the beast. One of the favorite chapters that Historicists love to employ in their teaching of historical 1260 YEARS, is Revelation 13. The argument of historicists is that if you believe that the first beast of Revelation 13 is Papacy (and by this they mean Medieval papacy), then you must believe that the 1260 days, if applied to it, can only make sense if it is 1260 years (day for a year principle). Consequently, you must admit that this started in 538 A.D. and ended in 1798 A.D.

But here is what bible students do not realize, and it is contained in the following statements,

"And I saw one of his heads as it were wounded to death; and his deadly wound was healed: and all the world wondered after the beast"(verse 3)

In this statement John was shown the Roman beast in its second revived Papal phase. Rome is to undergo transformations to prolong its life until the coming of Christ. The first transformation was to Medieval papal phase, which John saw as "WAS NOT YET IS". It is in this Medieval papal phase that the Roman beast sustained the mortal head wound. When it revives, it becomes the seventh head on the beast. It is this revived papal phase and the eighth head, not medieval papal phase, that is the main focus in Revelation 13. Therefore, it is the Revived Papal (seventh head) and the eighth head phase (rise from the bottomless pit) that the whole world wonders about and not the Medieval papacy. Of course, its medieval era papacy is cited, but done only briefly, so as to provide a historical perspective. The emphasis of Revelation 13 is on the very beast that Christ will destroy at his second coming, and this is the Revived Papal phase and the eighth head phase of the beast. It is this same beast that is described in Revelation 17 as the Scarlet Colored Beast and the Beast That Ascendeth from The Bottomless Pit, the Eighth Head, which is "Of the Seven"(i.e., one of the original seven heads in its second revived and evolved, more evil form). You will notice that also that in Revelation 17, the whole world is said to wonder after the beast, not in its first Imperial or medieval papal phase, but when it head wound is healed and in its final or eighth head rise from the bottomless pit, both of which are future. We read this in the following statements of John:

"The beast that thou sawest was, and is not; <u>AND SHALL ASCEND OUT OF THE BOTTOMLESS PIT</u>, and go into perdition: <u>AND THEY THAT DWELL ON THE EARTH SHALL WONDER</u>, whose names were not written in the book of life from the foundation of the world, when they behold the beast that was, and is not, and yet is".

Another clue that points clearly that the 1260 DAYS belong to the future Papacy can be seen in verse 5:

"And there was given unto him a mouth speaking great things and blasphemies; and power was given unto him **<u>TO CONTINUE FORTY AND TWO MONTHS</u>**. [6] And he opened his mouth in blasphemy against God, to blaspheme his name, and his tabernacle, and them that dwell in heaven. [7] And it was given unto him to make war with the saints, and to overcome them: and power was given him **<u>OVER ALL KINDREDS, AND TONGUES, AND NATIONS. [8] AND ALL THAT DWELL UPON THE EARTH</u>** shall worship him, whose names are not written in the book of life of the Lamb slain from the foundation of the world".

It will be noticed that the emphasized statements cannot apply to Medieval Papacy. The statement to CONTINUE FOR FORTY-TWO MONTHS cannot apply to Medieval Papacy because the Medieval phase was its first ever existence. The word CONTINUE has a clear meaning, which is that something can only continue when it has already begun previously. Notice also the word **ALL, "in all that dwell upon the earth", and "all kindreds, and tongues, and nations".** Medieval Papacy never had the absolute and complete dominance that these expressions indicate. In fact, most of the dark ages, Medieval Papal authority was challenged and checked by stubborn kings and Princes **of Europe. Then there is the Americas which was then an unknown continent. Neither did Rome's and Papacy's influence extend over the far eastern people of China, Korea, Mongolia, etc.**

Therefore the 1260 days applies, not to medieval era papacy, but instead to the revival of the Roman beast in the last days. It is this revived Roman beast that the whole world is to wonder about, some with admiration and some with dread (its revival and resurrection in America where it was most hated by the early puritan America is indeed a wonder and astonishment). America is now almost completely beholden to the Papacy. The early American pilgrims will roll in their graves if they should now see the influence and power of Papacy in America.

The 1260 days will therefore begin the day the Revived Papal Beast, with the help of American Theocratic Dictatorship, fully revives the eighth head form of the Roman beast, and have that MAN OF SIN ascend world political power. Moreover, if we focus on the phrase, "continue for forty-two months", this truth is further evident because continuation is predicated on something that has previously begun and had occurred prior. Thus, we see prophecy here saying that the forty-two months is a continuation of a previous rule of the Roman beast, in harmony with the truth about the two phases of the beast: **WAS** (Imperial), **IS NOT (**Lost its imperial form) **AND YET IS** (Medieval papal form **in which it sustained** deadly wound in 1798 A.D.), and **SHALL ASCEND OUT OF THE BOTTOMLESS PIT**(Its full imperial, god-less form is restored).

The 1260 days constitutes an important key to completing the puzzle constituted by the books of Daniel and Revelation. It connects almost all aspects of the two books, and any misinterpretation of this prophetic period guarantees that the events of these books can never be fully aligned, as God, through Spirit of Prophecy by his servant Ellen G. White, proclaimed to the church. It is also this inability to accurately align the events of these books that has delayed the latter rain.

The Jesuits in Seventh day Adventist church hierarchy will therefore continue to canonize all the historicist interpretations of the book of Daniel and Revelation, but especially the 1260, 1290, and 1335 days. Their purpose is that the church will not reach a perfect and complete understanding of the two prophetic books.

THE 1260 DAYS IN REVELATION 12

THE WAR IN HEAVEN: THE PROBLEM WITH THE HISTORIC INTERPRETATIONS OF THIS EVENT

The prophecy states that as soon as Satan, he goes after the Woman, and the woman is helped to flee into the wilderness for 1260 days. The case to be made here is that the war occurred way after the Medieval papal dark ages, which will mean that the 1260 days cannot be the presumed 1260 years of Medieval papacy.

The earliest, and most illogical, unreasonable, traditional view on this prophecy is that which presumes that this event happened even before Adam and Eve were created. Even a cursory read of John's description of this event will easily reveal that this traditional view makes no sense at all. So, why did our Millerite pioneers hold this view? It is puzzling to say the least. But let's look at how impossible it is for this war to have occurred before Adam and Eve were created.

"And when the dragon saw that he was cast unto the earth, he persecutes *THE WOMAN <u>WHICH BROUGHT FORTH</u> THE MAN CHILD*"-Revelation 12:

Notice the time difference between this event and that of Revelation 12:4, which reads as follows:

"And his tail drew the third part of the stars of heaven and did cast them to the earth: and the dragon stood before *THE WOMAN, WHICH <u>WAS READY TO BE DELIVERED</u>*, for to devour her child as soon as it was born".

In the first text, the WOMAN HAS ALREADY BECOME A MOTHER-SHE HAS ALREADY DELIVERED THE MALE CHILD.

But in the second text, the WOMAN IS NOT YET A MOTHER- SHE IS ABOUT TO DELIVER, GOING THROUGH THE BIRTH PANGS.

The chronology is unquestionable: the war in heaven takes place after, in fact way after, the event of the dragon dragging one third of the angels down to earth. The significance cannot be lost relative to the issue of when this war took place or is to take place as the case may be. The event of the Dragon using his tail to draw a third of the angels and casting them to the earth is much earlier and therefore clearly different from the event of the same Dragon itself being cast down, with his angels.

The first case was of the Dragon deceiving and drawing the angels he deceived down to the earth region where he subsequently deceived Adam and took the earthly kingdom from him. Satan decided that since they have lost favor with God and had no desire to live in heaven under God's terms, they should plan to take over a world where they can not only work in opposition to God but live on their own terms, contrary to the Law of God.

But after Jesus had come and spiritually defeated him and claimed back the right to be the king of the world when redemption is completed, the Dragon, with the angels he had drawn down to earth with his tails, went to heaven to attempt to dethrone God and Christ. What he could not win fairly, he decided to win by force. Consequently, and subsequently, they were defeated and cast down to earth.

Despite clear evidence to the contrary, historical teachers of Revelation 12 keep teaching that the war in heaven took place even before Adam and Eve were created. They refuse or are just incapable of seeing that the war in heaven is not the same as the event described earlier in the chapter in which the Dragon drew the third of the angels down to earth.

While Satan, of his own volition, left heaven and drawing with him the third of the angels he deceived, which is an event that happened just before or immediately after Adam and Eve were created, the war in heaven is instead a case where, instead of by his own volition, Satan was actually cast out, evicted by force. It is the condition of Satan in the aftermath of this war that Isaiah was shown in Isaiah 14:12-15.

To establish further that the war could not have occurred before Adam and Eve were created, let us read the following words of John:

"And the great dragon was cast out, that old serpent, called the Devil, and Satan, which deceiveth the whole world: he was cast out into the earth, and his angels were cast out with him"- Revelation 12:9.

Satan and his angels are cast out of heaven at a time when it could be true that he and his angels have been deceiving the whole world. If the war supposedly happened before even Adam and Eve were created, which world then have they been deceiving? Thus the absurdity of the historical take on this war is exposed.

Let's look at another important point. We read,

"Now is come salvation, and strength, and the kingdom of our God, and the power of his Christ: for the accuser of our brethren is cast down, which accused them before our God day and night. [11] And they overcame him by the blood of the Lamb, and by the word of their testimony; and they loved not their lives unto the death".

Note the word "NOW" because it is telling you when this war occurred. So, as a result of Jesus successfully executing this war against Satan and his angels and casting them out, SALVATION and STRENGTH, and KINGDOM OF GOD came to heaven at the time indicated by the angel as **"NOW".** The implication and fact cannot be mixed by any unbiased reader. This is profound, the fact that it was not until after this war that salvation and strength and kingdom of God could come to the heavenly inhabitants. But let us quickly establish the immediate point we seek, and that is that this war and the casting out of Satan and his angels happened ***at a time when it could be true that Satan and his angels have been accusing the saints of God on a daily basis, BEFORE GOD, that is in the very presence of God in heaven.***

Think about this for a moment, no day passes without Satan accusing the saints before God, which means no day passed that he did not have access to and appeared before the presence of God accusing one saint or the other. Can you imagine the degree of annoyance this constituted to God and the other heavenly inhabitants? Can you imagine the lack of peace and tranquility this constituted heaven. This completely refutes the presumption that heaven is at peace and all blissful, and restful. Fact is, since Satan rebelled and added Adam's rebellion to his, heaven has never been at peace, not until Satan and his angels are cast out. This is what this war is about. It is only when this war is concluded that the Angel could declare, **"NOW IS COME SALVATION (TO HEAVEN, NOT TO EARTH YET) ..."**

The story of Job reveals only one example of what Satan has been doing to every saint. It would be naïve to think that Job's case was the only case. Granted his case stands out, which is why God included it in scripture. The truth is that God simply chose Job's experience as the one He will put in the scriptures. There is no single saint that will be saved that Satan will not bring accusation to God about. And until the day this war is fought in heaven and he and his angels are cast out, Satan will be making these accusations in person in heaven and before the throne of God, to God's face. This is the indignation that God and the whole heaven have been enduring for the sake of our salvation.

Was Satan's accusations limited to only the saints of the old testament era? Again, it would be naïve to believe that. Ask yourself, what was Satan doing in heaven when Jesus made the following statement,

John 12:31-33 KJVS

"Now is the judgment of this world: now shall the prince of this world be cast out. [32] And I, if I be lifted up from the earth, will draw all men unto me. [33] This he said, signifying what death he should die".

"I beheld Satan as lightening, fall from heaven." (Luke 10:18).

How can Satan accuse the saints, and leave off accusing their Master Jesus Christ while he was here on earth accomplishing our salvation? Yes, there was no day that passed while Jesus walked our earth that Satan was not in heaven accusing him before Jehovah his father. In the above statements, Jesus was prophesying what will happen when his ministry was completed, and he ascends back to heaven.

Jesus's successful ministry judged and condemned Satan. The judgement sentence was passed on Satan. But the casting down of Satan also happened in the real time that he made the above statement, just that it was not the final casting out. God presented to Christ a temporary demonstration of what he will do later on a permanent basis during the war John saw and recorded in Revelation 12. At this juncture, we can see that there is no way the war John described in Revelation 12:7-10 could have occurred before Adam and Eve were created.

But what about the other equally historical view that this war occurred when Jesus ascended back to heaven and was glorified, after his ministry on earth was completed. After discovering how untenable the view of PRE-CREATION WAR was, some of these historical teachers evolved to the second view, namely, that the war must have occurred when Christ went back to heaven after his earthly ministry. With this later view, they hope to still justify the teaching that the 1260 DAYS belonged to Medieval Papacy.

But is this view scriptural, is it tenable? Let's find out.

John wrote:

"Now is come salvation, and strength, and the kingdom of our God, and the power of his Christ: for the accuser of our brethren is cast down, which accused them before our God day and night. [11] ***And they overcame him by the blood of the Lamb, and by the word of their testimony; and they loved not their lives unto the death"***.

Note verse 11 carefully. Who do you think could be described as those who overcame by the blood of the Lamb?

Can you see that the view that this war occurred as soon as Christ ascended back to heaven completely excludes saints of the New Testament era. Can this thought even be fathomable? If there are people who it could be truly, in more direct sense, said that they overcame by the blood of the Lamb, is it not the saints of the era after Christ's actual death and actual, literal shedding of his blood?

Does it make sense to suggest that Satan left out the saints of the New Testament era for accusation before God? Even if you claim that, well Satan did not need to go to heaven to accuse the saints of the New Testament era, how can you explain excluding these saints from the class that ***overcame Satan by the blood of Jesus and loved not their lives unto death?*** Impossible.

Therefore, this war that John saw did not occur with the death of Jesus or as soon as he ascended to heaven and was glorified. This war must therefore occur at such a time when it could be said that Satan has accused every saint that will be in heaven and lost his case against them. This cannot be until the day that God has numbered and determined ALL REMAINING SAINTS ON EARTH THAT WILL BE SAVED, just before Jesus come, in other words ALL THE SAINTS OF THE LAST GENERATION must have been accused by Satan. Then, with no one else to accuse, and having lost case not only against Christ but against those he has advocated for and saved,

Satan will be, for the FINAL TIME, cast out of heaven and into the earth, permanently barred from heaven, and his EVIL FOOTPRINTS IN HEAVEN forever ERASED(CLEANSED).

Satan and his angels are the FIRST CASUALTIES OF THE WORK OF CLEANSING THE HEAVENLY SANCTUARY.

Let us recap what we have learned so far.

Revelation 12:10-11 KJVS

"And I heard a loud voice saying in heaven, NOW IS (HAS) COME SALVATION, and strength, and the kingdom of our God, and the power of his Christ: for the accuser of our brethren is cast down, which accused them before our God day and night".

"[11] And they overcame him by THE BLOOD OF THE LAMB, and by the word of their testimony; and they loved not their lives unto the death".

It is so hard for me to believe that verse 11 refers only to saints before the CRUCIFIXION.

By interpreting this war to have been fulfilled when Jesus was crucified, saints of the era after the cross, including the apostles of Jesus who were with him on earth, are excluded from those whom Satan had been accusing and who overcame by the blood of the Lamb, and were martyred for their faith in Jesus.

Let us realize that until Christ died, the blood of the Lamb had not literally been shed. How then can the saints that yielded their lives in the faith of Jesus after the blood had been actually shed be excluded? This to me constitutes the major flaw of the view that Revelation 12:7-12 was fulfilled at the time of the crucifixion.

Therefore, we must look for the fulfillment of this prophecy at a later time from the "CROSS", a time which can make sense of the phrase,

"because he knoweth that he hath but A SHORT TIME"(verse 12).

DETERMININATION OF WHEN THE WAR WILL OCCUR AND THE FINAL CASTING OUT OF SATAN WILL HAPPEN: THE CONNECTION WITH THE 1290 PLUS 1260 DAYS OF PERSECUTION

An important clue comes from what the Devil will do as soon as he is cast out and into the earth,

"And when the Dragon saw that he was cast unto the earth, he persecuted the woman which brought forth the man child. And to the woman were given two wings of a great eagle, that she might fly into the wilderness, into her place, where she is nourished for a time, times, and half a time, from the face of the serpent"- Revelation 12:13,14.

AT THIS JUNCTURE THEREFORE, WE ESTABLISH THAT SATAN WAS, OR WILL BE CAST DOWN, AS THE CASE MAY BE, JUST A SHORT TIME CLOSE TO OR BEFORE THE BEGINNING OF THE 1260 DAYS. Whether this is future or historic depends therefore on whether the 1260 DAYS is historic or still future. Since this cannot be determined conclusively in the context of Revelation 12, it requires the examination of the other texts and contexts where the 1260 DAYS are mentioned ***Daniel 12, Revelation 11 and 13***.

So far in this study, we have shown from the examination of the other places the 1260 DAYS are mentioned that this prophetic period is literal and future. With those preponderance of evidence and proof, we can conclude also, with confidence, that the 1260 DAYS here in Revelation 12, is also 1260 literal days that are to be fulfilled in future. Therefore, we conclude that the war in heaven will occur just before the beginning of the 1290 + 1260(2550 DAYS or 7 years) of theocratic American and Revived Papal Rome's persecution of the saints. Soon after this war in heaven, and Satan sees that he is now cast out into the earth and never to escape the confines of the earth, he will move upon the leaders of the world, religious as well as political, to legislate the Sunday Law and other activities that will constitute the TAKING AWAY OF THE DAILY that Daniel saw (Daniel 12). Then the last wave of persecution and killing of the saints, which God mentioned in the fifth seal, commences. Then, 1290 days later God takes the persecuted church into the wilderness to prepare her to stand during the seven trumpets and the seven last plagues without spot and wrinkle, and ready for immortality when their Master and Savior comes.

With this we have exhaustively examined all the messages in which the 1260 DAYS are mentioned and proved that in no instance, occasion or context, do they point to 1260 YEARS that is already fulfilled and past. Rather they all consistently point to literal days that are yet to be fulfilled in the near future from the present year of 2023 A.D.

Now that we understand that these periods are literal and future, we can easily understand why they are given and how we can arrange them, which Daniel could not, because more knowledge about the subject was to come from the books of the New testament, especially the book of Revelation written by John.

THE ARRANGEMENT OF DANIEL 12 PROPHETIC PERIODS AND THE URGENT LAST GENERATION MESSAGE THEY HAVE FOR THE CHURCH

A cursory look at these three prophetic periods may make it seem as though they are disjointed. But when one realizes that they are given as an answer to a question, then it should be evident to the reader that there is an order and sequence to them. One important key to understanding the sequence and order to these three prophetic periods, is to assign them to two main categories, and these are:

1. **PERIOD(S) OF PERSECUTION OF THE SAINTS ("DANIEL'S PEOPLE").**
2. **PERIOD OF DELIVERANCE-BLESSING (RESURRECTION AND IMMORTALIZING OF THESE SAINTS).**

That the taking away of the always Daily marks the commencement of persecution for the saints is evident both in Daniel 8 as well as Daniel 11:30-35. Also the prophetic period of the 1260 days(42 months or "time, times and an half), in all the places it is mentioned, is associated with the persecution by the Roman beast. Based on this knowledge, the reader will quickly see that two of the three prophetic periods are periods during which the saints will undergo persecution in the hand of this power Daniel had been seeing and have been describing with the pronoun "HE", which is the same power described as ABOMINATION OF DESOLATION. This power is clearly Rome in all its phases of rulership-Imperia(fifth head)l, Medieval papal(Sixth head), Revived

Papal(Seventh head), and the last(Eighth head) phases. These two prophetic periods appointed for the persecution of the saints are the 1290 DAYS and 1260 DAYS. On the other hand, the prophetic period of the 1335 DAYS is associated with and appointed for the BLESSING OF THE SAINTS. In the context of Daniel 12, the BLESSEDNESS alluded to here, is clearly that of the STANDING UP OF MICHAEL (Jesus Christ) TO DELIVER THE EMBATTLED LIVING SAINTS, THE RESURRECTION OF THE DEAD SAINTS, and the IMMORTALITY OF ALL THE SAINTS, and these in turn relate to the SECOND COMING OF JESUS CHRIST, because He is the one that will bring about these blessings when He comes.

Thus, it becomes evident that these prophetic periods are anything but disjointed answers to the question: "How long shall it be to the end of these wonders?" But what are the DAILY and ABOMINATION OF DESOLATION? In the next section we will examine the church's traditional views on the subjects of the DAILY and ABOMINATION OF DESOLATION, to prove that they are inconsistent and out of harmony with plain understandings of scripture.

THE DOCTRINE OF THE "DAILY"

THE "DAILY" AND "TAKING AWAY OF THE DAILY" IN THE ERAS OF THE JEWISH, DARK AGES, AND LAST DAYS CHURCHES

Why does SEVENTH-DAY ADVENTISTS CHURCH preach only one fulfillment of the *TAKING AWAY OF THE DAILY*- the one she interprets as *PAGANISM TAKEN AWAY BY PAPACY IN 508 A.D. BY CLOVIS*? Is the church's interpretation of the *DAILY* and *TAKING AWAY OF THE DAILY*, correct? How many occurrences of *TAKING AWAY THE DAILY* did Daniel see? Could it be that Seventh-day Adventist church is making the same mistake about the *DAILY*, which William Miller made regarding the SANCTUARY, and consequently positioning herself for another major disappointment when Jesus does not come as early as they expect?

These are the questions that will be answered in this study.

DAILY AND ABOMINATION OF DESOLATION ARE LIKE TWO OPPOSITE SIDES OF A COIN-MUTUALLY EXCLUSIVE TO EACH OTHER AND AT ENMITY AGAINST EACH OTHER

Just as the two sides of a coin cannot be facing up or facing down at the same time, so are the DAILY and ABOMINATION OF DESOLATION. While one is facing up, the other one must be facing down. They are not only parallel to each other but are moving in opposite directions and so can never meet. They are MUTUALLY EXCLUSIVE, meaning that they cannot and do not COEXIST in the same place, the place in question being THE SANCTUARY. The two have nothing and hold nothing in common. Yet, when you see Abomination set up and in place, the Daily is not very far away because the latter always precede the former. The Daily has to be taken away from the sanctuary or holy place in order for the abomination of desolation to be st up in its place. This fact is clearly seen in Daniel 11.

"And arms shall stand on his part, and they shall POLLUTE THE SANCTUARY OF STRENGTH, and shall take away THE DAILY sacrifice, and they shall place the ABOMINATION THAT MAKETH DESOLATE"- Daniel 11:31 KJVS

Clearly, it can be seen that the Daily has to be taken away from the sanctuary first before the Abomination of desolation can be set up in the sanctuary. Since these are spiritual actions, as evident from the words ***POLLUTION***, and ***HOLY COVENANT*** in the previous verse, we are therefore talking about **TWO OPPOSING SPIRITUAL PHENOMENA AND POWERS.** We can now establish the important point and fact that Abomination of desolation does not come, is not set up, except the Daily is taking away first, and therefore the two are not very far away from each other. This being the case, why is it not understood by the church that there was a ***DAILY*** **VERSUS** ***ABOMINATION OF DESOLATION DYNAMICS or CONTROVERSY*** that also occurred in the Jewish church era, specifically in relation to JESUS' EARTHLY MINISTRY? Jesus was very clear that ***ABOMINATION OF DESOLATION***(AOD) was associated with the Jewish nation-church, as we can read in the following texts:

"When ye therefore shall see the abomination of desolation, spoken of by Daniel the prophet, stand in the HOLY PLACE, (whoso readeth, let him understand:)"- Matthew 24:15 KJVS

"But when ye shall see the abomination of desolation, spoken of by Daniel the prophet, standing <u>WHERE IT OUGHT NOT</u>, (let him that readeth understand,)" Mark 13:14 KJVS

"And when ye shall see JERUSALEM compassed with armies, then know that the desolation thereof is nigh.[24] And they shall fall by the edge of the sword and shall be led away captive into all nations: and Jerusalem shall be trodden down of the Gentiles, until the times of the Gentiles be fulfilled".

QUESTION: What does the phrase, "where it ought not" mean? Luke 21:20-24 KJVS

ANSWER: Where did it not originally belong.

QUESTION: What originally belongs there?

ANSWER: The "Daily". Hence the need to remove or take away the Daily first before Abomination of desolation can occupy.

All these texts establish that Abomination of desolation was set up at the time Jesus was on earth, and a few years later came to **STAND (IMPOSE ITSELF)** in the holy place (Jerusalem and the sanctuary) and destroy it. Knowing that this does not happen unless the DAILY IS TAKEN AWAY first, we therefore have the alert that the TAKING AWAY OF THE DAILY must have occurred just prior to the coming of the Abomination of Desolation. The question therefore is: When was the Daily taking away in the days of Christ. Who and What is the Daily in the time Jesus was on earth? Ministers read the above texts on Abomination of desolation **(AOD),** as it relates to the Jewish church and nation, but cannot understand that Abomination of Desolation is always associated, in an opposing, mutually exclusive relationship, with the **DAILY**. They talk about three, and some talk about four occurrences of Abomination of desolation, but they only teach just one and only occurrence of the **DAILY BEING TAKEN AWAY**?

THE DAILY IN THE JEWISH ERA AS SEEN BY DANIEL IN DANIEL 8

THE DAILY IS THE DEFAULT OCCUPANT OF THE SANCTUARY: When something is the default, it means it is the first and original component or service of the manufactured product or system. While using the equipment, users can either remove the default service and select their own or customize the equipment according to their preferences. This is the analogy that explains the dynamics of the DAILY and the ABOMINATION OF DESOLATION.

God's sanctuary, whether the earthly or the heavenly, always have the DAILY as the default component or occupant or "software". The sanctuary runs DAILY as originally. It is the daily that makes the sanctuary HOLY, and RIGHTEOUS, and therefore fittingly GOD'S HOUSE. Abomination of desolation is nowhere near, not to talk of being part of the sanctuary of God. Therefore, for Abomination of desolation to be **SET UP**, it requires that the DAILY be removed or ***TAKEN AWAY*** first. This is because it is the position and place of the Daily which the Abomination of desolation has to occupy.

So, while the Daily is present in the "HOLY PLACE" (or SANCTUARY OF STRENGTH), Abomination of desolation does not have a place in there. It therefore means that when you see the Abomination of desolation in the sanctuary, the Daily was there a short time earlier and had been taken away. How then can teachers talk about FOUR ABOMINATION OF DESOLATION that will manifest in the history of the church, but mention only ONE DAILY episode? The answer is because they are ignorant of what the Daily is and the SPIRITUAL DYNAMICS that exist between those two opposite and opposing powers.

Later, we will see how this answers Paul's question in 2 Thessalonians 2, about ***HE WHO NOW LETTETH WILL LET UNTIL HE BE TAKEN OUT OF THE WAY***. Paul is talking about "THE DAILY" being the one that will "**LET**", i.e., PREVENT, until he be taken out of the way. The phrase "taken out of the way" clearly indicate forceful and violent removal. When we have identified WHO and WHAT the DAILY is, Paul's message will become very clear.

On this note, let us go ahead and identify THE DAILY.

IDENTITY AND IDENTIFICATION OF THE DAILY AND WHAT IT MEANS TO TAKE IT AWAY FROM THE SANCTUARY

DAILY AND ABOMINATION OF DESOLATION COMPETE FOR THE SAME PLACE, STATUS, AND AUTHORITY IN THE SANCTUARY ("HOLY PLACE"): We have established that wherever you find Abomination of desolation, it means that the DAILY was there earlier and had to be taken away. This, as we have shown, is evident from Daniel 11:31. Therefore, the fact that Jesus saw in the book of Daniel, that Abomination of desolation is associated with the Jewish church and nation, means that he and Daniel equally saw the DAILY and the TAKING AWAY OF THE DAILY in relation to the Jewish church and nation. This means therefore that the Daniel 11:31 DAILY-ABOMINATION OF DESOLATION controversy is the second case instead of the one and only case, as SEVENTH-DAY ADVENTISTS CHURCH falsely teaches. In fact, we will see in this study that there are altogether THREE OCCURENCES OF THE TAKING AWAY OF THE DAILY that Daniel saw, with the Daniel 12 occurrence being the THIRD, LAST, and still FUTURE, case that is yet to be fulfilled. Some teachers make the case for four Abomination of desolations by including the Babylonian captivity example. Well then, the case can also be made for a fourth taking away of the Daily. But Daniel never mentioned the Abomination of desolation in relation to the Babylonian destruction of Jerusalem, and so teachers should not presume to do so. In this study, we will therefore restrict ourselves to what Daniel was given.

WHAT IS THIS DAILY IN JESUS'S TIME-AS SEEN BY DANIEL IN DANIEL 8?

Between Daniel and Jesus' prophecies, we see that there was a Jewish church era DAILY VERSUS ABOMINATION OF DESOLATION CONTROVERSY that began in Jesus' days and concluded in A.D. 70.

This first case of the controversy was recorded in Daniel 8, as we read:

Daniel 8:11-12 KJVS

"Yea, he magnified himself even to the prince of the host, and by him the daily sacrifice was taken away, and the place of his SANCTUARY was cast down. [12] And an host was given him against the daily sacrifice by reason of transgression, and it cast down the truth to the ground; and it PRACTISED and PROSPERED".

Looking at this passage, it can be noticed that the little horn (Rome) makes attack against Prince of the host (Jesus), the Truth (Jesus), and the Daily. Based on the principle of parallelism, the Daily has to be Jesus Christ also and the concept of his truth, plan of salvation and sanctuary ministrations. Based on this accurate interpretation, the message we have from the above passage is about the crucifixion of Christ by Rome with the alliance of Jewish leaders(host), who as a result of forsaking the covenant and law of God, became the ally of Rome in this wickedness.

SO, WHOSE SANCTUARY IS IT?

In the context of Daniel 11:30-33, we get a corroboration of the deduction and conclusion made so far.

"And arms shall stand on his part, and they shall pollute the sanctuary of strength, and shall take away the daily sacrifice, and they shall place the abomination that maketh desolate"- Daniel 11:31 KJVS

The first thing that stands out or should stand out to the reader and need to be focused on, is the word SANCTUARY, because it is so central to this DAILY VERSUS ABOMINATION OF DESOLATION controversy. This is because it is the place where this DAILY is to be found, and from where it's taken away, that the Abomination of desolation is set up to takes its place. Note the words ***POLLUTE***: it is an action that is committed against the ***"SANCTUARY OF STRENGTH".*** It spiritually connotes what is done against that which is **SACRED** and **HOLY**. It can be noticed here that it is the act of **TAKING AWAY THE DAILY** and **SETTING UP THE ABOMINATION OF DESOLATION** that brings about the pollution of the sanctuary of strength. It is how the sanctuary of strength is polluted or defiled.

Based on traditional SEVENTH DAY ADVENTISTS teaching on the DAILY-ABOMINATION OF DESOLATION subject, the Daily is identified as ROMAN PAGANISM, which makes the ***SANCTUARY OF STRENGTH*** to be Roman, the Roman Pantheon, as some have suggested. So, in essence and in fact, ADVENTIST ministers are of the view that Roman pagan shrine, with its pagan worship, is a HOLY and SACRED PLACE, that it can be polluted, and that Abomination of desolation (PAPAL ROME) is to remove this pollution? ***Talk of Devil casting out Devil.*** Is that even possible? This case of NAKED ROMAN PAGANISM being taken away by DISGUISED ROMAN PAGANISM is what ADVENTIST ministers believe as the interpretation of Daniel 11:31?

This interpretation, as can be seen, is an absurdity. It negates all rules of interpretation of prophecy. The only difference between NAKED ROMAN PAGANISM and PAPAL ROMAN PAGANISM is merely political. Otherwise, they are the same spiritually, the same "God"-the SUN-TRINITY GOD, the same ceremonies, the same DAY OF WORSHIP, and of the same origin, which is Satan. There cannot be any controversy between the two. At best, one will simply yield and cede to the other peaceably, which is what both Emperors Constantin and Justinian did in ceding HEADSHIP OF THE CHURCH to the Pope of Rome. They simply made their own kind, Pope or Bishop of Rome, head of the church, whom they know is simply a **CAESAR IN SACERDOTAL CLOTH**.

But it is clear that what Daniel saw in Daniel 11:30-33, is not political and leadership difference and opposition. It is **CONTROVERSY BETWEEN GOOD AND EVIL, BETWEEN GOD AND SATAN**. This is supported even more when we read the previous verse 30 and the verse after:

Daniel 11:30 KJVS

"For the ships of Chittim shall come against him: therefore, he shall be grieved, and return, and have ***INDIGNATION AGAINST THE HOLY COVENANT:*** so shall he do; he shall even return, and have intelligence with ***THEM THAT FORSAKE the HOLY COVENANT***".

Daniel 11:32 KJVS

"***And such as <u>do wickedly against the covenant</u> shall he corrupt by flatteries***: **<u>BUT</u>** the ***people that do know their God*** **shall be strong and do exploits".**

The two spiritually opposing sides are clearly revealed; those who forsake or do wickedly against the holy covenant versus those that keep and honor the holy covenant, with the former going into alliance with this king or power, the Roman power.

In these two texts, we have a context that shows that the ***SANCTUARY OF STRENGTH*** is not some Roman shrine or worship place, but rather ***GOD'S SANCTUARY*** (the ***HOLY PLACE*** **OF GOD AND JESUS HIS HIGH PRIEST**), which is why Jesus used the expression, "stand where it ought not. If Papal Roman power occupies the place where Ancient Pagan Rome initially use to be, can it be rightly said of papacy that it is occupying where ***IT OUGHT NOT TO***, or that it is occupying ***THE HOLY PLACE***? What can be holy about Ancient Roman shrines and temples of worship?

Therefore, if the taking away of the daily and placing of Abomination of desolation in its place constitutes the pollution of this sanctuary of strength, it means that **THE DAILY is of God and ABOMINATION OF DESOLATION is of Satan.** This is the theme and subject of **THE DAILY VERSUS ABOMINATION OF DESOLATION CONTROVERSY, spiritual battle between Good and Evil, between Righteousness and Wickedness.** It is therefore this same theme and subject that Daniel was shown in Daniel 8:10-12, which is the first occurrence of this controversy that Daniel was shown in vision.

IDENTITY OF THE DAILY: THE MANUAL OR SOFTWARE OF THE SANCTUARY

Jesus declared, "I am the WAY, the TRUTH, and the LIFE; no man cometh unto the Father, but by me" (.

The Psalmist confessed, "Thy WAY, O God, is in the SANCTUARY…Thou art the God that doest wonders: thou hast declared thy STRENGTH'' (Psalms 77:13,14).

This is in a nutshell what the sanctuary, its articles, and its ministration are about, the salvation of humankind and bringing them to God the Father. Jesus is saying that he is the only one appointed and commissioned by Jehovah to accomplish this task. In the above words, Jesus summarizes his high priestly work in the sanctuary, **TO ADMINISTER THE WAY OF GOD, THE TRUTH OF GOD, AND THE LIFE OF GOD to the WORSHIPPERS IN THE SANCTUARY, referred to as the HOST OF THE SANCTUARY.** By just answering the question, *Who originally occupies the holy sanctuary*?, we have ourselves the identity of the Daily, which is that ***THE DAILY IS GOD THE FATHER (AND JESUS HIS SON AND HIGH PRIEST) PLUS THE HOLY COVENANT AND ITS ADMINISTRATION.*** The DAILY is therefore the engine on which the sanctuary runs, using automotive analogy. Without the daily the sanctuary is not spiritually functional and becomes an empty shell. This therefore explains why Jesus declared to Jews:

"Behold, now your house is left unto you desolate".

THE GRAND CONCLUSION OF THE IDENTITY OF THE DAILY is therefore that ***THE DAILY IS THE OCCUPANT(S) OF THE SANCTUARY, THE "PROGRAM" THAT IS RUN OR ADMINISTERED IN THE SANCTUARY, AND THE METHOD OR FORMAT THE ADMINISTRATION FOLLOWS.***

With the understanding that **THE DAILY IS OF GOD**, while **ABOMINATION OF DESOLATION** (ROMAN POWER AND GOD) **IS OF THE DEVIL, and that the two are opposite and mutually exclusive, it means that when Abomination of desolation takes the place of the daily in the sanctuary, SATAN, AND HIS SON, WHICH THE HUMAN REPRESENTIVE HE HAS CHOSEN are in place, and the program and format changes. Instead of the DOCTRINES OF GOD AND JESUS CHRIST, it becomes the doctrines of the Devil and his Human son that id being run in the sanctuary. When Rome came in A.D.66, it was Satan and Nero his son with the sun God that came to Jerusalem and the purpose is to officially impose the worship of the sun God (which is Satan) on the nation. This is the interpretation of Daniel 8:10-12.**

Daniel 8:10-12 KJVS

"And IT waxed great, even to the host of heaven; and it cast down some of the host and of the stars to the ground and stamped upon them. [11] Yea, he magnified himself even to the PRINCE OF THE HOST, and by HIM (enemy of the Daily and Prince of the host) THE DAILY sacrifice WAS TAKEN AWAY, and the PLACE OF HIS (Prince of the host and daily's) SANCTUARY was cast down. [12] And an host was given HIM against the daily sacrifice by reason of transgression, and IT cast down THE TRUTH to the ground; and IT practised, and prospered".

We already know that this "***IT***" power is Ancient Rome because we know that it is the power that came after the fall of Greece, and most importantly we know that it is the power that ***directly magnified itself against Christ***, the Prince of the host, by crucifying him. But the identification that is very crucial here, is that which

Jesus gave it when he identified Rome as the ***Abomination of Desolation***. Here are Jesus' words:

""And when ye shall see JERUSALEM compassed with ARMIES, then know that the desolation thereof is nigh.[24] And they shall fall by the edge of the sword and shall be led away captive into all nations: and Jerusalem shall be trodden down of the Gentiles, until the times of the Gentiles be fulfilled"- Luke 21:20-24.

We know whose armies encompassed and destroyed Jerusalem, the Roman armies, and that in Matthew's and Mark's record of the same prophecy, this power is called the ABOMINATION OF DESOLATION. With this definition and identification of the Little horn power, let us paraphrase the above texts. We will have the following:

"And [ABOMINATION OF DESOLATION] waxed great, even to the host of heaven; and it cast down some of the host and of the stars to the ground and stamped upon them. [11] Yea, [ABOMINATION OF DESOLATION] magnified himself even to the *PRINCE OF THE HOST*, and by ABOMINATION OF DESOLATION *THE DAILY* sacrifice WAS TAKEN AWAY, and the PLACE OF HIS (PRINCE OF THE HOST'S) SANCTUARY was cast down. [12] And an host was given ABOMINATION OF DESOLATION *AGAINST THE DAILY* sacrifice by reason of transgression (JEWISH CHURCH APOSTACY), and ABOMINATION OF DESOLATION cast down *THE TRUTH* to the ground; and it practised, and prospered".

How so clearly this prophecy can now be understood. This is exactly how Jesus read this prophecy, which enabled him to see that even though Daniel did not use the exact words ***ABOMINATION***

OF DESOLATION, it was strongly and clearly implied and indicated. It shows the parties in the Great Controversy clearly and reveals which party is of God and which is of the Devil. Evidently, ***THE DAILY BELONGS TO PRINCE OF THE HOST (JESUS CHRIST) AND THEREFORE IS OF GOD***, while ***LITTLE HORN*** (or Rome), identified by Christ as the ***ABOMINATION OF DESOLATION***, belongs to the devil obviously.

So, we know who is to ***TAKE AWAY THE DAILY***, and that it is the ***ABOMINATION OF DESOLATION***, and he has to do this because it wants to take the place of the ***DAILY*** in the sanctuary. He gets help from those who transgress the holy covenant, **"an host" ("by reason of transgression"),** which is the apostate Jewish leaders and followers. **This apostate "host", as can be seen, is in contrast from the "host" that has Jesus as their Prince.** ***By reason of transgression***, that is as a result of **FORSAKING THE HOLY COVENANT and WORSHIP OF JEHOVAH,** this apostate Jewish leaders and church constituted themselves as allies or host of the Abomination of desolation (Rome), against Jesus, and together they crucified him thereby **CASTING THE TRUTH PERSONIFIED(JESUS) TO THE GROUND AND PROSPERED and REPLACING HIS TRUE WORSHIP for THE HONOR, AUTHORITY AND WORSHIP OF CAESAR AND SATAN.**

This is exactly what happened later when this prophecy fulfilled, because the Jewish nation and leadership declared, in their wicked rejection of Jesus and the Holy New Covenant that he came to confirm among them, **"We have no king but Caesar".** It is thus, with these words, that they rejected JEHOVAH(AND JESUS), the **KING OF THEIR NATION AND HEAD OF THEIR CHURCH**, and in his place openly accepted Caesar as the King of their nation and Head of their church (We can see how the drama here is similar to the one Daniel recorded in Daniel 11:30-32, where the second occurrence of the TAKING AWAY OF THE DAILY AND SETTING UP OF ABOMINATION OF DESOLATION was fulfilled), which led to the **POPE(Caesar in sacerdotal cloth)**, being christened the head of the church in the place of her true Head and High Priest, Jesus Christ, and thus again the honor, authority, worship of God and the priesthood of His Son cast to the ground, and replaced by Roman and Caesar's priesthood.

At this juncture, it is evident who and what the Daily is because as we can see, it is clearly in connection with the Sanctuary, and therefore in connection with the Truth, the Way, and the Life of God, as ministered in the sanctuary. We therefore ask: Who is the embodiment of the sanctuary and its services and its important goal, which is ETERNAL LIFE AND SALVATION OF THE HOSTS OF THE SANCTUARY? Who is the embodiment of the Truth, the Way, and the Life of God? The answer is very obvious, it is God the Father (and Jesus Christ), to whom every sanctuary article and those ***DAILY***(CONTINUAL) ***AND YEARLY*** ministrations pointed to.

A second and complementary meaning of the DAILY, comes from realizing the WHOM that the sanctuary is built for, the ONE who is to inhabit it, the ONE whose presence is to ***CONTINUALLY***("DAILY") abide in it. He is the one that is the ***DAILY*** **(or CONTINUAL)**. And this is none other than God the Father, Jehovah. What did God tell Moses? - ***"Let them build me a sanctuary that I may dwell among them".*** This dwelling in the sanctuary by God, was to be on a "***DAILY***" basis.

All together therefore, we can conclusively say that the ***DAILY IS JESUS, AND BY THE UNION BETWEEN FATHER AND SON, THE DAILY IS ALSO GOD THE FATHER***. It is the two of them that those ***DAILY MINISTRATIONS AND SERVICES OF THE SANCTUARY*** represented, and in those ***DAILY SERVICES*** it is God and Jesus that were the object of the ***WORSHIP, PRAISES***

AND PRAYERS OF REPENTANCE, CONFESSIONS, AND FORGIVENESS. It is ***THEIR PRESENCE*** that made the sanctuary ***HOLY and RIGHTEOUS***. Thus when THEY (THE DAILY) are TAKEN AWAY, the sanctuary became SPIRITUALLY DESOLATE, as Jesus told them in Matthew 23:

"Behold your house is left unto you desolate".

And when this happens, the next inevitable event is that the spiritually desolate soon becomes physically desolated, as happened in A.D.70. They then constitute themselves as allies with the enemy of God and His son. By this **UNHOLY, ABOMINABLE UNION (TRANSGRESSION OF DESOLATION),** what we describe today with the terms, ***CHURCH, AND STATE UNION***, which is also BABYLON THE GREAT IN THE UNION OF SPIRITUAL FORNICATION WITH THE STATE, they spiritually, and where possible, physically, crucify God and His son. The TAKING AWAY OF THE DAILY therefore means, and happens, when God's professed, but apostate, people reject the Holy Covenant and worship of God, and instead accept and anoint the devil, and whoever his human representative is at the time, thereby making that human representative of Satan to be HEAD OF HIS CHURCH and their KING and RECIPIENT OF THEIR honor and worship. This always lead to the destruction of the apostate institution that had rejected God. In the last days, we read that the TEN NATIONS will destroy Babylon and burn it with fire(Revelation 17 and 18), in fulfillment of God's plan and judgement against this apostate body that used to have His name.

THE REJECTION AND CRUCIFIXION OF CHRIST WAS THE FIRST OF THREE OCCURRENCES OF TAKING AWAY THE DAILY-<u>A.D. 31.</u>

Jesus (and by association God the Father), as THE DAILY PERSONIFIED, was taken away from the sanctuary, and the result is what Jesus expressed in the following words:

"**Behold, your HOUSE** (THE SANCTUARYAND THE CITY AND NATION) **is left unto you** [SPIRITUALLY] **DESOLATE**"-Matthew 24:38.

No sooner had Jesus had made this declaration, and they openly declared, **"We have no king but Caesar"-**John 19:15**,** thus rejecting and replacing the Daily with the Abomination of Desolation-CAESAR OF ROME AND HIS SUN GOD. Not knowing it then, they had judged and condemned themselves and the nation to the hand of the very Power that will physically desolate them in a few short years, as Christ and Daniel prophesied.

We thus have a pattern of events delineated by Daniel 8-9 and Matthew 24, as follows:

1. **THE PROFESSED CHURCH AND PEOPLE OF GOD FORSAKES THE HOLY COVENANT.**

2. **THE APOSTATE PEOPLE THEN TAKE AWAY THE DAILY FROM THE SANCTUARY AND SET UP ABOMINATION THAT MAKETH DESOLATE IN THE DAILY'S PLACE.**

3. **THE ABOMINATION OF DESOLATION POWER THEN COMES AND TAKES A STAND TO DEMAND ABSOLUTE WORSHIP AT THE PENALTY OF DEATH AND DESTRUCTION TO THE PEOPLE AND OR NATION THAT REFUSES TO OBEY AND WORSHIP IT.**

This pattern was to be repeated two more times, in answer to the question asked in Daniel 8:13:

"How long shall be the vision concerning the daily sacrifice, and the transgression of desolation, to give both the sanctuary and the host to be trodden under foot"

The answer that was given was as follows:

"Unto two thousand and three hundred days; then shall the sanctuary be cleansed."

Thus, from the answer given, we learn that the above pattern and sequence of events are to continue and repeat until the sanctuary is cleansed, and then this power that desolates, treads underfoot the sanctuary, the daily and the truth and the saints of the sanctuary, will forever be destroyed, when the kingdom of this world becomes the kingdom of our Lord and of His Christ, as John saw when the seventh trumpet began to sound.

Until the sanctuary is cleansed, the treading underfoot of the sanctuary and host will continue, but at the end of the 2300 days (October 22, 1844), Christ began a program to end the power of this Abomination of Desolation, to ensure that when it comes to power the third time, it will be the last, and it will not again **PROSPER** and **PRACTISE** (i.e., **LONGEVITY**) as it did in the first and second occurrences.

THE SECOND OCCURRENCE, as we have shown, is the one that occurred and came to a head when Emperor Justinian completed what Emperor Constantine began in 325 A.D., by ultimately making the Pope (Caesar in sacerdotal cloth) the King and Head of the Churches, in the place of Jesus the son of God. This happened in 533 A.D.

THE THIRD AND LAST OCCURRENCE OF TAKING AWAY THE DAILY AND SETTING UP ABOMINATION OF DESOLATION-DANIEL 12-MAKING AN IMAGE TO THE BEAST-REVELATION 13

In furtherance of the question and answer that began in Daniel 8:13,14, heavenly beings engaged in another question-and-answer drama for Daniel's and of course the church's benefit. We read as follows:

Daniel 12:5-7 KJVS

Then I Daniel looked, and, behold, there stood other two, the one on this side of the bank of the river, and the other on that side of the bank of the river. [6] And one said to the man clothed in linen, which was upon the waters of the river, How long shall it be to the end of these wonders? [7] And I heard the man clothed in linen, which was upon e the waters of the river, when he held up his right hand and his left hand unto heaven, and sware by him that liveth for ever that it shall be for a TIME, TIMES, AND AN HALF; and when he shall f have accomplished to scatter the power of the holy people, all these things shall be finished.

Daniel 12:8 KJVS

And I heard, but I understood not: then said I, O my Lord, what shall be the end of these things.

Daniel 12:9-13 KJVS

And he said, Go thy way, Daniel: for the words are closed up and sealed till the time of the end. [10] Many shall be purified, and made white, and tried; but the wicked shall do wickedly:

and none of the wicked shall understand; but the wise shall understand.

[11] And from the time that THE DAILY SACRIFICE SHALL BE TAKEN AWAY, AND THE ABOMINATION THAT MAKETH DESOLATE SET UP, there shall be a thousand two hundred and ninety days.

[12] Blessed is he that waiteth, and cometh to the thousand three hundred and five and thirty days. [13] But go thou thy way till the end be: for thou shalt rest, and stand in thy lot at the end of the days.

OBJECTION TO THE "DAILY" IN DANIEL 8:11,12 BEING RELATED TO THE JEWISH SANCTUARY

Let us address an important objection by those who cannot see the sanctuary in Daniel 8:11,12 as referring to the earthly sanctuary. They point out that by A.D. 70 when the sanctuary was destroyed, it had ceased to be God's and Christ's sanctuary, and the city had ceased to be God's and Jesus' holy city. So, in their mind, Daniel 8:11,12 must be referring, not earthly sanctuary, but to either Roman Pagan Sanctuary or may be Heavenly sanctuary. But let these critics realize that in Jesus's prophecy about the destruction of Jerusalem, an event that occurred in A.D. 70, 39 years after the crucifixion, he still referred to it as the ***HOLY PLACE***, and as ***A PLACE WHERE ABOMINATION OF DESOLATION OUGHT NOT TO STAND***:

"And when ye shall see JERUSALEM compassed with ARMIES, then know that the desolation thereof is nigh.[24] And they shall fall by the edge of the sword and shall be led away captive into all nations: and JERUSALEM shall be trodden down of the Gentiles, UNTIL the times of the Gentiles be fulfilled"- Luke 21:20-24.

"When ye therefore shall see the abomination of desolation, spoken of by Daniel the prophet, stand in the <u>HOLY PLACE,</u> (whoso readeth, let him understand) …"

"But when ye shall see the abomination of desolation, spoken of by Daniel the prophet, standing <u>WHERE IT OUGHT NOT</u> (let him that readeth understand).

So, as can be seen, just because Jerusalem and the earthly sanctuary ceased to be God's spiritually and officially, does not mean that it has lost that **HISTORICAL** (not spiritual) significance and mention as God's holy sanctuary and city. It is simply a matter of fact and recognition of history that the Angel made in Daniel 8:11,12, in the same way that Jesus made it in his prophecy about the destruction of Jerusalem and the sanctuary. **Historically**, Jerusalem will never cease to be ***THE PLACE OF GOD'S AND JESUS' SANCTUARY***. It is in this sense that Angel Gabriel made the statement in the vision to Daniel.

Therefore, the accurate interpretation of the **DAILY**, is that it represents God in union with His son Jesus, and their Truth, Way, and Life (or the Gospel of Salvation) and worship, that are being accomplished in the sanctuary and administered in the spirit of believers. When they are removed by the union of apostate church and state, and abomination of desolation is set up in their place, the sanctuary of God and its apostate host become SPIRITUALLY TRODDEN UNDERFOOT and DESOLATE.

Usually, shortly after that, the Abomination of desolation power is permitted by God to STAND UP AND PHYSICALLY DESOLATE the Apostate church or people. This is exactly what happened to the Jewish nation-church, as has been explained earlier in the study, and was the first occurrence, to be followed by two more occurrences before Jesus will put an end to it with the

CLEANSING OF THE SANCTUARY, followed by the seven last plagues and defeat of this evil power in the war of Armageddon.

GRAND CONCLUSION

Now that we know who and what the Daily is and what taking it, and or HIM, away means, we can now see how and where Ellen G. White's 4000, 6000 years (or the 2000 years factor) connects THE TWO LAST GENERATIONS. While Daniel 8 and 9 provide the events and dates of the LAST GENERATION OF THE JEWISH NATION, Daniel 12 provides the parallel, equivalent events for the LAST GENERATION OF THE JEWISH NATION.

The two last generations share two events in common, namely, TAKING AWAY OF THE DAILY and ABOMINATION OF DESOLATION SET UP AND STANDING IN THE HOLY PLACE. Of these two, Ellen G. White's 4000, 6000 years (i.e. the 2000 years) provides the link between the TWO EVENTS OF TAKING AWAY OF THE DAILY. Since the first one climatically occurred in A.D. 31 in the crucifixion of Jesus and their declaration that Caesar is their King, 2000 years later being 2031, will be the year when Daniel 12 equivalent event of taking away of the daily will occur.

Thus, Ellen G. White's prophetic messages of the 4000, 6000 years provides one of the two crucial pieces of information Daniel sought for, namely, how to link the end of the 2300 DAYS with Daniel 12 prophetic periods. With the proper arrangement of the prophetic periods of Danie 12, as we have done earlier in the study, we can now correctly come up with the year Jesus will come to be 2038 A.D.

This conclusion can best be visualized in the charts shown below.

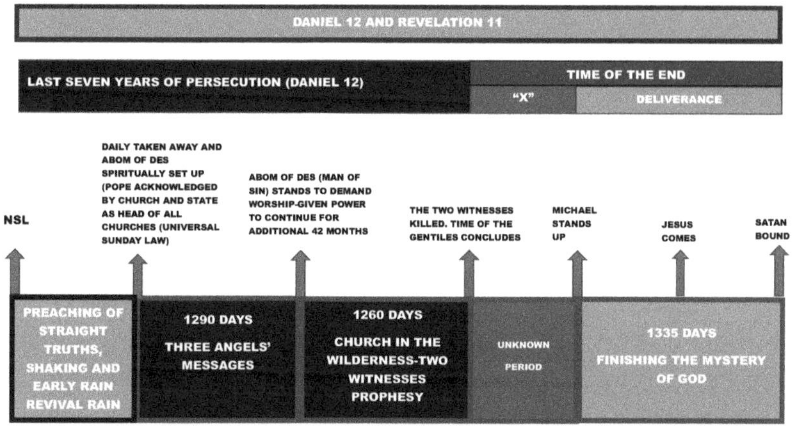

UNKNOWN PERIOD: The period between the end of the sixth trumpet and the sounding of the seventh trumpet (Revelation 11:14,15).

NSL: America first passes the Sunday law (make an image to the beast), and for an unknown period of time will campaign and persuade the rest of the nations of the world to do the same.

USL: At this point the complete image of the beast would have been formed, and severe persecution will commence for God's people. Image of the beast will speak and cause that whoever will not worship it will be killed. No one will buy or sell except they have the mark, the name, or the number of the beast. This is the last period for God's people to proclaim the three angels' messages in power of the latter rain. At the end of this period, God will give His people two wings of an eagle (the same way He gave Philip) to flee into the wilderness (all the desolate and hidden places of this earth where they will be nourished both physically and spiritually.

NOTICE that because of the unknown period, it is impossible to know the day and hour of Jesus coming, unless God reveals it to the 144000, thus Jesus' declaration that no one will know the day and hour before hand attested to.

SPIRIT OF PROPHECY STATEMENTS ON 4000, 600 YEARS

"*Satan led men into sin, and the plan of redemption was put in operation. For four thousand years, Christ was working for man's uplifting, and Satan for his ruin and degradation. And the heavenly universe beheld it all. When Jesus came into the world, Satan's power was turned against Him*"(DA page 759).

"*For four thousand years the race had been decreasing in physical strength, in mental power, and in moral worth; and Christ took upon Him the infirmities of degenerate humanity. Only thus could He rescue man from the lowest depths of his degradation*"(DA page 1675)

"*It would have been an almost infinite humiliation for the Son of God to take man's nature, even when Adam stood in his innocence in Eden. But Jesus accepted humanity when the race had been weakened by four thousand years of sin. Like every child of Adam, He accepted the results of the working of the great law of heredity. What these results were, is shown in the history of His earthly ancestors. He came with such a heredity to share our sorrows and temptations, and to give us the example of a sinless life*"(DA page 756)

www.ingramcontent.com/pod-product-compliance
Lightning Source LLC
LaVergne TN
LVHW050915210926
848482LV00012B/445
9798295577802